Key Skills Leve

Communication; Application of Number; Information and Communication Technology

Written to the 2004 Standards

Roslyn Whitley Willis and Liam Gabrielle

Series Editor

Roslyn Whitley Willis

Published by
Lexden Publishing Ltd
www.lexden-publishing.co.uk

To ensure that your book is up to date visit:
www.lexden-publishing.co.uk/keyskills/update.htm

Acknowledgements

The authors wish to acknowledge Alison Bell for her contribution and ideas made in the early stages to this book.

Thanks go to my husband for his contribution in naming some of the fictitious organisations that, and people who, appear in this book, together with his forbearance and encouragement.

I wish to dedicate this book to my father Peter Whitley – *Roslyn Whitley Willis.*

Special thanks for their tireless support and patience go to my wife and family and to my co-author, Roslyn, who was always willing to give direction in this mammoth task from the very start of an idea – *Liam Gabrielle.*

We would both like to thank the publisher, Mark Kench, whose belief and hard work helped the project reach fruition.

First Published in 2006 by Lexden Publishing Ltd.

Cover photograph of juggling balls by kind permission of Marcel Hol ©

British Library Cataloguing in Publication Data.

A CIP record of this book is available from the British Library.

ISBN-10: 1-904995-12-8

ISBN-13: 978-1904995-12-8

Typeset and designed by Lexden Publishing Ltd

Printed in Malta by Gutenberg Press.

Lexden Publishing Ltd
23 Irvine Road
Colchester
Essex CO3 3TS

Telephone: 01206 533164
Email: info@lexden-publishing.co.uk
www.lexden-publishing.co.uk

Preface

The material in this book gives you the opportunity to understand Key Skills and practise them so you are able to meet the high standards set out in the Level 2 Key Skills Standards for:

- Communication;

- Application of Number; and

- Information and Communication Technology.

The introductory section of this book explains each of the Key Skills and how to gain a qualification.

The book is then divided into three separate chapters covering the three Key Skills mentioned above.

Each chapter is further divided into three distinct parts:

1   Reference Sheets

These sections provide all the necessary background information to prepare you for each of the Key Skills. They provide useful exercises that will:

- aid your learning;

- can be used for revision; and

- prepare and aid you for the Part A Tasks and End Assessment questions.

2   Part A Practice Tasks

Working through these will help you produce work at the right level and prepare you for the End Assessment.

As you complete each task you will become more confident about what is expected in Key Skills and be able to use your knowledge and understanding to pass the End Assessment and put together a Portfolio of Evidence.

3   End Assessment Questions

These sections provide examples of the type of questions that are likely to appear on an End Assessment paper and that you may have to pass as part of your Key Skills qualification.

Further resources

When your tutor thinks you have enough knowledge of Key Skills, she/he will give you an assignment, or assignments, to complete. Working successfully through the assignment(s) will show you are able to apply your knowledge and understanding, and produce work that will go into your Key Skills Portfolio of Evidence. These assignments are contained in the *Tutor's Resource* cd.

Additional resources and information can be found at www.lexden-publishing.co.uk/keyskills.

Software

The software referred to throughout this publication is Microsoft Office 2003® and Internet Explorer 6® running on Microsoft Windows XP®.

Contents

WHAT ARE KEY SKILLS? – A STUDENT'S GUIDE

Key Skills are important for everything you do, at school, at college, at work and at home. They will help you in your vocational studies and prepare you for the skills you will use in education and training and the work you will do in the future.

Key Skills are at the centre of your learning, and the work in this book provides you with the opportunity to develop and practise the Key Skills of Communication, Application of Number and Information and Communication Technology, through a variety of tasks. Having Key Skills knowledge will help you apply them to other areas of your studies.

There are six Key Skills

Communication is about writing and speaking.	**Application of Number** is about numbers.	**Information and Communication Technology** is about communicating using IT.
It will help you develop your skills in: ⚫ speaking; ⚫ listening; ⚫ researching; ⚫ reading; ⚫ writing; ⚫ presenting information in the form of text and images, including diagrams, charts and graphs.	It will help you develop your skills in: ⚫ collecting information; ⚫ carrying out calculations; ⚫ understanding the results of your calculations; ⚫ presenting your findings in a variety of ways, such as graphs and diagrams.	It will help you develop your skills in using computers to: ⚫ find and store information; ⚫ produce information using text and images and numbers; ⚫ develop your presentation of documents; ⚫ communicate information to other people.
Improving Own Learning and Performance is about planning and reviewing your work.	**Problem Solving** is about understanding and solving problems.	**Working with Others** is about working effectively with other people and giving support to them.
It will help you develop your skills in: ⚫ setting targets; ⚫ setting deadlines; ⚫ following your action plan of targets and deadlines; ⚫ reviewing your progress; ⚫ reviewing your achievements; ⚫ identifying your strengths and weaknesses.	It will help you develop your skills in: ⚫ identifying the problem; ⚫ coming up with solutions to the problem; ⚫ selecting ways of tackling the problem; ⚫ planning what you need to do to solve the problem; ⚫ following your plan; ⚫ deciding if you have solved the problem; ⚫ reviewing your problem solving techniques.	It will help you develop your skills in: ⚫ working with another, or several, person(s); ⚫ deciding on the roles and responsibilities of each person; ⚫ putting together an action plan of targets and responsibilities; ⚫ carrying out your responsibilities; ⚫ supporting other members of the group; ⚫ reviewing progress; ⚫ reviewing your achievements; ⚫ identifying the strengths and weaknesses of working with other people.

HOW TO GAIN A KEY SKILLS QUALIFICATION

Mandatory Key Skills

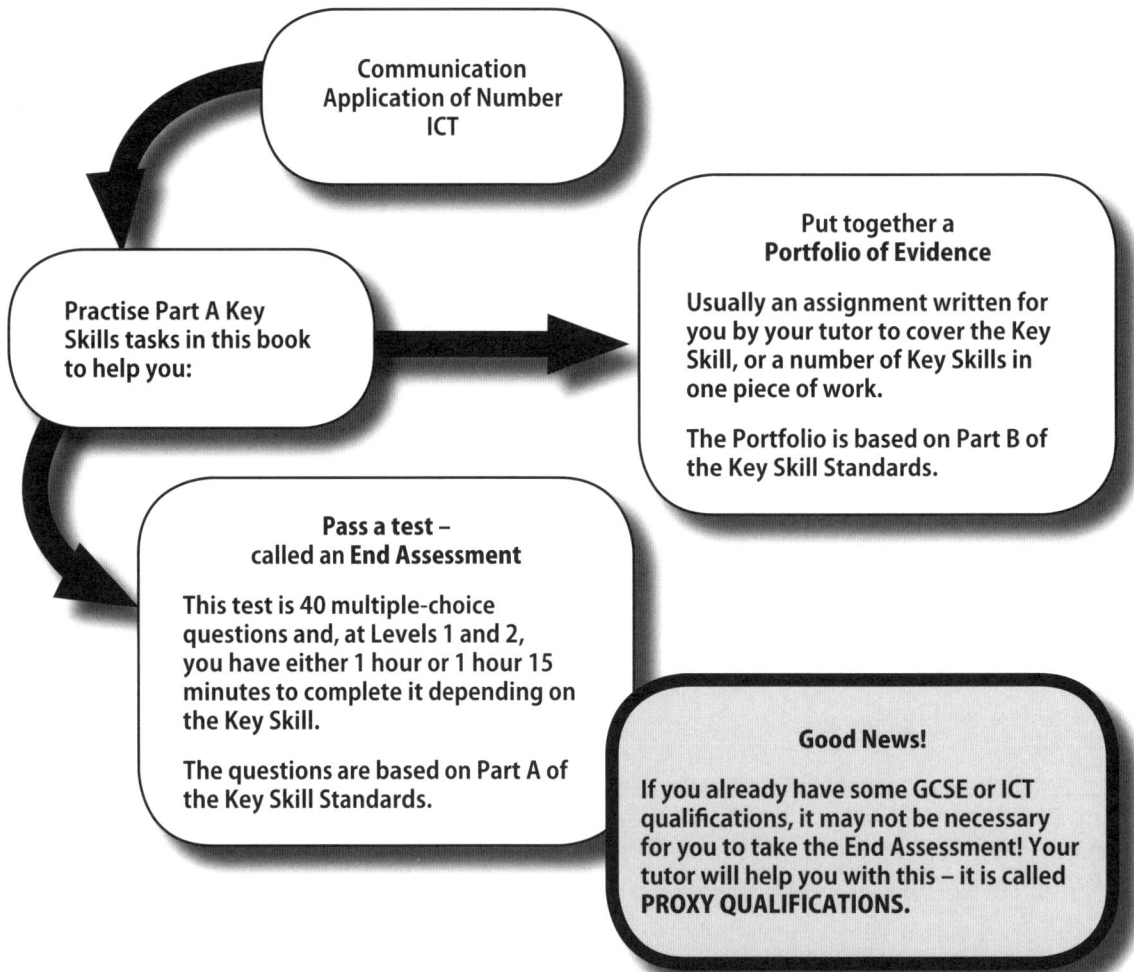

> Communication
> Application of Number
> ICT

> Practise Part A Key Skills tasks in this book to help you:

> **Put together a Portfolio of Evidence**
>
> Usually an assignment written for you by your tutor to cover the Key Skill, or a number of Key Skills in one piece of work.
>
> The Portfolio is based on Part B of the Key Skill Standards.

> **Pass a test – called an End Assessment**
>
> This test is 40 multiple-choice questions and, at Levels 1 and 2, you have either 1 hour or 1 hour 15 minutes to complete it depending on the Key Skill.
>
> The questions are based on Part A of the Key Skill Standards.

> **Good News!**
>
> If you already have some GCSE or ICT qualifications, it may not be necessary for you to take the End Assessment! Your tutor will help you with this – it is called **PROXY QUALIFICATIONS.**

Wider Key Skills

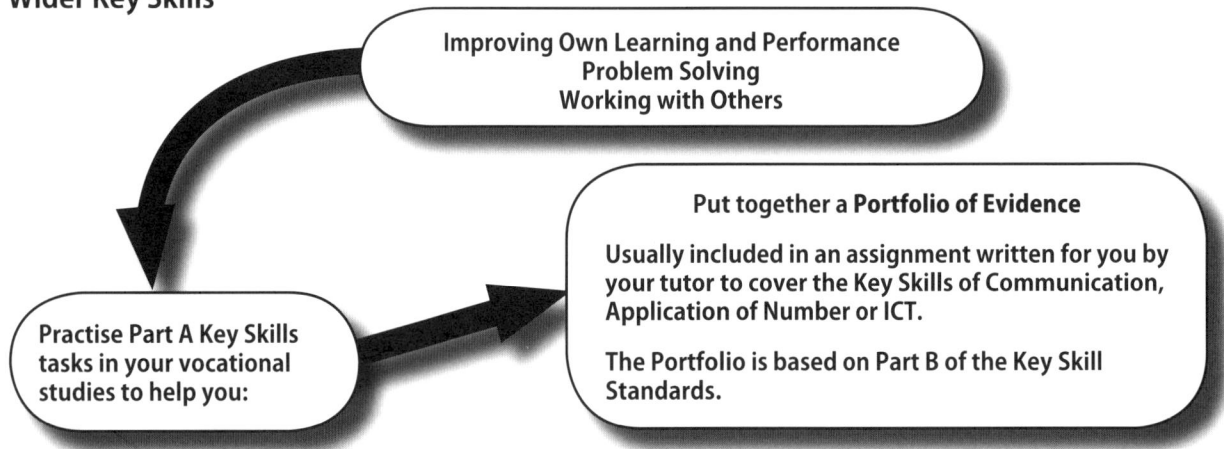

> Improving Own Learning and Performance
> Problem Solving
> Working with Others

> **Put together a Portfolio of Evidence**
>
> Usually included in an assignment written for you by your tutor to cover the Key Skills of Communication, Application of Number or ICT.
>
> The Portfolio is based on Part B of the Key Skill Standards.

> Practise Part A Key Skills tasks in your vocational studies to help you:

Opportunities to work towards achieving the Wider Key Skills are provided in the Portfolio assignment work and are included in the *Tutor's Resource* that accompanies this text.

THE PORTFOLIO

STEP 1

Once your tutor has assessed your assignment work and you have passed, you will put your work into your portfolio.

A **Portfolio of Evidence** usually takes the form of a lever arch file with a **Portfolio Front Sheet** that shows:

- where you are studying;
- which course you are studying;
- which Key Skill(s) are in the portfolio;
- when you passed your End Assessment(s); and
- details of any Proxy Qualifications.

STEP 2

It is important to number every page of the work you put in your portfolio. This helps you complete the **Log Book** that your tutor will give you.

STEP 3

Complete the Log Book. This indicates where your evidence is to be found and also describes what is in the portfolio.

STEP 4

Check your Log Book entries carefully, making sure everything is correct and neat.

Get your tutor to check you have put your Portfolio together correctly.

STEP 5

Sign the Log Book and get the person who assessed your work to sign too.

Once you have completed your Portfolio of Evidence it is shown to someone outside your centre whose job it is to check it meets the Key Skills Standards. If this person agrees that it does, then you have **passed your Portfolio of Evidence**.

Chapter 1: Communication

At **Level 2**, learners need to use the skills of speaking, listening, reading and writing. You will be able to take part in discussions using a varied vocabulary and help to move the discussion forward so that it flows freely and allows everyone to contribute.

Learners will be able to select reading material from a variety of sources. Such material will contain up to 500 words and will be on a variety of topics . You will show you can summarise this material, correctly following the meaning of the documents. You will be able to write documents, some of them up to 500 words in length, to suit the purpose of the task, and the audience who will read it. You will include relevant images in some of the documents you write. Your work will be correctly spelt and punctuated and you will use correct grammar. Learners will give a short talk that will last at least four minutes.

The following Reference Sheets provide opportunities for you to review and practise the Communication skills needed for Key Skills.

WRITING AND SETTING OUT MEMOS

A memorandum – plural memoranda
(abbreviated to memo)

A memo is an **internal** method of communication.

Memos must be short documents, and usually deal with one subject. A long document within an organisation is usually sent in the form of a report.

The memo should be signed by the sender.

Although organisations have their own style of layout for memos, all memos contain these essential headings:

MEMORANDUM

Mrs A Winston, Personnel Manager

T Gilbert, Central Records Manager

15 June 2006

Lost file

Mr J Brown, Personnel Director
Miss P Patty, Central Records Clerk

To
From
Date
Subject
Copies to

The subject of the memo has been identified.

This section indicates who else, other than the named recipient, has received a copy.

Last week I informed you that Mrs Jane McTavish's file had been lost or mislaid.

I am pleased to report that this has now been found and I have written to Mrs McTavish apologising for the delay in confirming the details she requested.

I am sorry for the inconvenience this has caused all parties.

Trevor Gilbert

Trevor Gilbert

Typical layout of a memorandum (memo). This is formal as it includes their titles (Mr, Mrs, Personnel Manager, etc.).

In this example you will see the message is short and simple and deals with only one point.

Who the memo is from, and to whom it is being sent, are identified and the document is dated and signed.

Informal memo

This is an example of an **informal** memo:

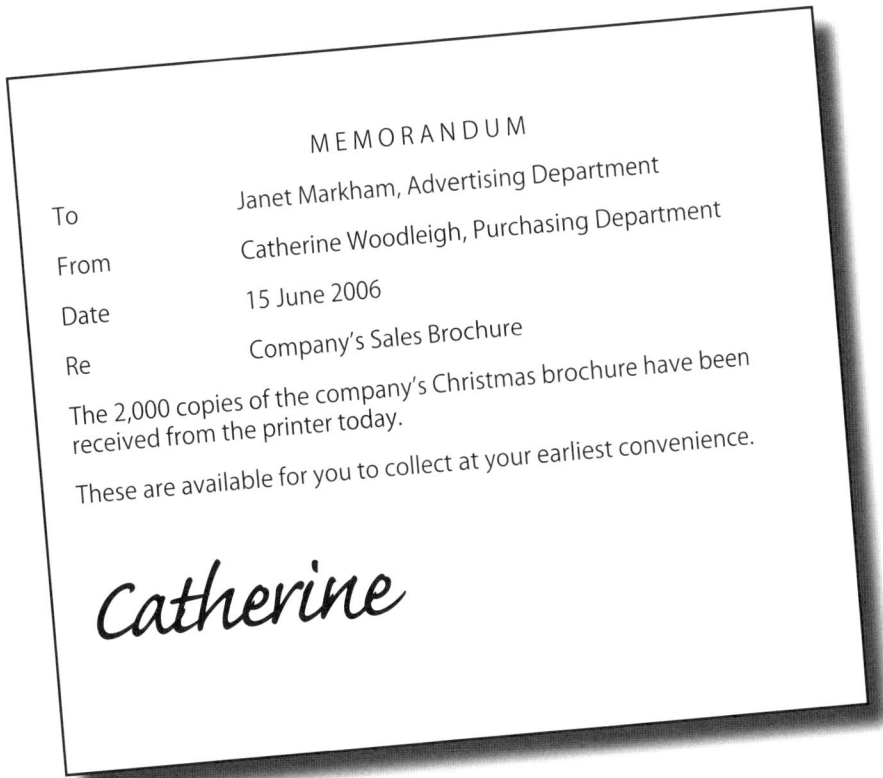

MEMORANDUM

To	Janet Markham, Advertising Department
From	Catherine Woodleigh, Purchasing Department
Date	15 June 2006
Re	Company's Sales Brochure

The 2,000 copies of the company's Christmas brochure have been received from the printer today.

These are available for you to collect at your earliest convenience.

Catherine

In this example, the names have no title and there are no job titles included – although people's departments are shown. This is important because an organisation could employ people with the same name but who work in different departments.

The memo is dated. There is a subject – this time expressed as 're' (short for 'reference').

The memo is signed with the sender's first name only – the surname could be included.

TAKING MESSAGES

It is always helpful to use a standard message form to record a message, whether it is a telephone message or a message of another kind. The headings on a standard form will help you include the information needed.

There is a message form on *page 9*, but this is not the only layout that companies use.

Remember you should always:

- Use simple, straightforward words.

- Keep your sentences short but vary the length a little so that the message reads well.

- Include **all, and only, the key facts and information.**

- Leave out irrelevant information.

- If you are repeating a request for the reader of the message to **do something**, make it a request **not an order**.

- Be very specific and clear about days, dates and times. If you have to give a non-specific time, e.g. "tomorrow", add the day and date in case your message is not read immediately. It is advisable to always be specific about days and dates in order to avoid confusion.

- Mark urgent messages clearly.

Your responsibility does not end when you place the message on the right desk – it only ends when the person has read it and understood it.

Identifying the key facts

Every message contains key facts. If you miss them out of the message it will not make sense – or not make **complete sense**.

Business callers are normally quite good at giving the key facts in an ordered way and checking them through afterwards. Private callers may be less helpful and some may like to chat, so that it becomes difficult to sort out what is important from what is not.

A good way to check you have the message clear in your mind is to read back your summary to the caller. This both checks that you have the message correctly with all the important facts, and gives the caller the opportunity to alter or add anything.

> **Note**
>
> This page includes the word **message**. This is a commonly misspelt word that you should be familiar with and learn to spell correctly. We have provided a document of commonly misspelt words that can be downloaded at www.lexden-publishing.co.uk/keyskills or ask your tutor for a copy. Use this to learn the words that cause you problems.

MESSAGE FORM

TO..DEPARTMENT...

DATE ....TIME ....

CALLER'S NAME ....

ORGANISATION ....

TELEPHONE NUMBER...FAX NUMBER ....

EMAIL ADDRESS ....

✓ Appropriate box(es)

Telephoned	☐
Returned your call	☐
Called to see you	☐
Left a message	☐
Requests you call back	☐
Please arrange an appointment	☐

Message

....

....

....

....

....

....

....

Taken by... Department... Time..............................

USING IMAGES IN COMMUNICATION

Images can be used to enhance and explain written communications.

Remember, use images to **enhance the text**, and to help the reader's understanding of the text. An image may also provide information in addition to text. **An image should not be included if it has no relevance**.

Think carefully about why you are using images and only use appropriate images in appropriate places.

Presenting numerical data in visual form

There are a number of situations when you will find it necessary, or preferable, to produce visual representations of numbers. Some people find it easier to understand figures when presented in graphical form, rather than table form. By all means consider using both a table and a graph, thus providing a number of ways in which the reader can understand the information.

Data in tabular form

INTERNATIONAL TEMPERATURES		
3rd January 2006		
CITY	MIN	MAX
Lisbon	10	14
Madrid	-2	12
London	3	12
Brussels	6	10
Amsterdam	6	8
Helsinki	1	2

Data in graphical form

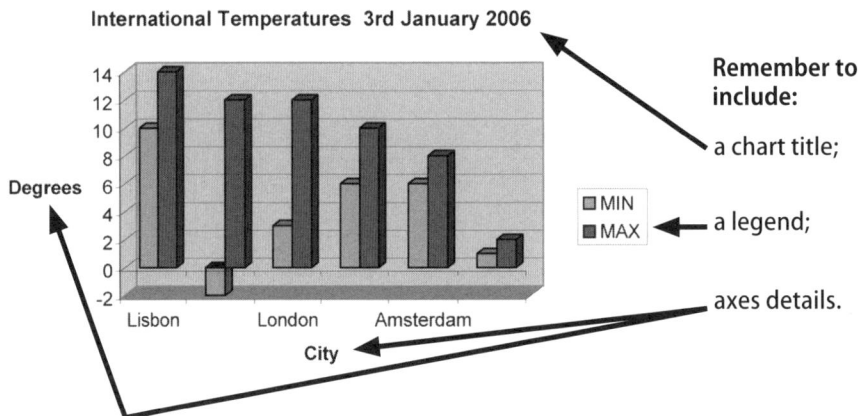

Remember to include:

a chart title;

a legend;

axes details.

The purpose of including the chart and the table is to allow the reader to understand and interpret the information in the most suitable way.

CRUISING ON THE AIDAblu
– a P&O* Cruise Liner

Imagine... it's 7.30 am and the sun is just beginning to rise above the mountains that run down to the sea... you've had breakfast in one of the four restaurants... you are on deck watching the beautiful island of Madeira get closer as your floating, luxury hotel edges slowly into port... the forecast for the day ahead is 28°... you've all day to explore today's destination and will probably have dinner in one of the restaurants as AIDAblu leaves port at 8 pm.

Come and experience the relaxing life on board our latest cruise liner

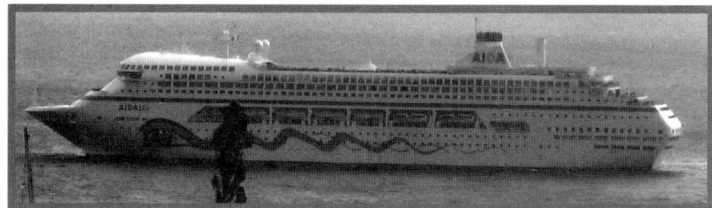

The AIDAblu entering the port of Funchal, Madeira

We'll include visits to some of the most beautiful islands and ports in the Atlantic

*A member of Cruise Ports World

The inclusion of an image in this article helps the reader to identify the 'product' that is being discussed and adds interest to facts and figures.

In this instance, the *text* is being used to aid the readers' understanding of the images.

WEATHER UK
20th February 2006

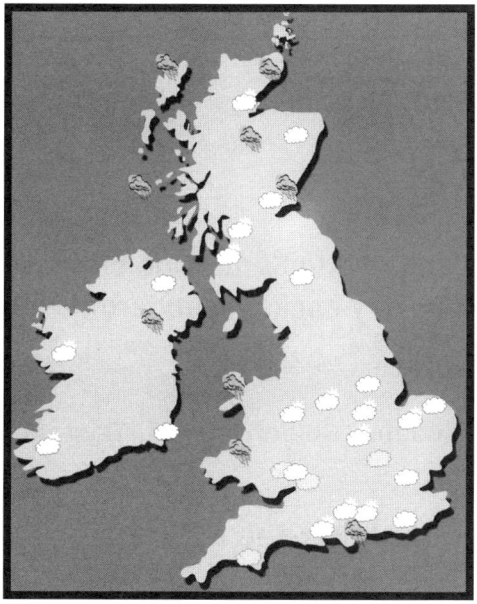

**WEATHER UK
20th February 2006**

Norfolk and Suffolk

There are few clouds at 2000 feet

Visibility is 7000m

Cornwall and Devon

Light rain at 1300 feet

Broken rain clouds at 1600 feet

Wind speed 18-36 mph

ADVERTISEMENTS

Advertisements may be placed in newspapers or magazines for a number of reasons, including:

- ✓ to advertise jobs;
- ✓ to promote products or services;
- ✓ to announce special events or functions;
- ✓ to publicise changes in an organisation;
- ✓ to recall faulty goods.

The **Classified Advertisements** section of a newspaper allows quick reference to a wide range of advertisements which are usually inserted according to subject.

Line advertisements

> GOOD BUY, BRAND new telephone/fax/ copier/scanner for sale. Owner is relocating abroad. Tel: 0184 576399

This information runs from line-to-line, often using the same typeface throughout, with no special layout. Charges are made by the line, normally with a minimum charge for three or four lines.

In such advertisements (also know as lineage ads), an opening should be made which catches the readers' attention, and then as much abbreviated information as possible should be contained in as few lines as possible.

Display advertisements

These may use a variety of fonts and sizes, and may be illustrated with artwork and colour. Charges are based on the number of column centimetres, often with a minimum size. Information can be displayed within the advertisement to attract attention to special features.

Porto Santo Lines

£35

ONE DAY CRUISE

Discover a new island. Porto Santo Line offers you an unforgettable one-day-package, aboard the ship "Lobo Marinho".

Travel with us and find out why Porto Santo is called the "Golden Island". Contact us today to make your reservation.

Porto Santo Lines, Rua da Praia 6, Funchal, Madeira Tel/Fax 291 228 662

Column advertisements (in newspapers and magazines)

The pages of newspapers and magazines are divided into **columns** and advertisers purchase so many column widths. The publisher charges so much per column and depth of advertisement.

In this example the page has been divided into four columns. Hop, Skip and Jump has taken an advertisement over two columns.

HOP, SKIP AND JUMP

Shoe manufacturers of quality

<u>END OF SEASON SALE</u>

Leather Uppers ● Leather Soles
● Luxury Comfort Linings

Sizes 5, 6, 7, 8, 9, 10 and 11.

Brogue	Black	Brogue	Brown
Oxford	Black	Oxford	Brown
Casual	Black	Casual	Brown
Lace	Black	Lace	Brown

Telephone to place an order TODAY

Whilst stocks last

0165 7873 9882

Designing advertisements

Designing an advertisement is an exercise in **summarising**. It is important to pick out the main points, features, advantages, or whatever is relevant to the theme of the advertisement.

It is essential to ensure the advertisement will be **seen** on the page of the newspaper or magazine. If it is displayed unattractively, it will not achieve this objective. Here are some guidelines:

- Use a company logo, prominently displayed. People can identify with a well-known logo.

- Whatever is being advertised, display the headline **PROMINENTLY** using bold text, underlining, **ALL CAPITALS**, for instance.

- Break up the information sensibly and logically; perhaps various points could be listed using an asterisk or a bullet point.

- Use spacing and balance sensibly – remember the more space you use the more you will pay!

- Try to achieve a progressive display which categorises information logically, leading finally to action required by the reader – "visit us on ???" "Contact us", etc.

1

WRITING AND SETTING OUT BUSINESS LETTERS

A business letter is an **external** method of communication and reflects how an organisation communicates with, and is viewed by, people and organisations outside the business.

There are a number of purposes for business letters:

- providing information;
- giving instructions;
- confirming arrangements;
- improving customer services;
- public relations.

A business letter has three parts:

1   introductory paragraph;

2   middle paragraph(s);

3   closing paragraph.

Introductory paragraph

The introduction/opening paragraph introduces the theme/purpose of the letter and puts it into a context or provides a background.

Introductory paragraphs are also used to mention essential people, events or things to which the letter will refer.

Middle paragraph(s)

These provide detailed information.

The middle paragraphs of a letter **develop a theme** and **provide all relevant details** and particulars. The number of paragraphs used will depend upon the complexity of the letter's subject. However, paragraphs should be kept fairly short and deal with only one topic at a time. **New topic = new paragraph** is something you must keep in mind.

Closing paragraph

This provides an action statement and a courteous close.

In this paragraph you will attempt to summarise your comments and state what action you will take, or wish to be taken.

Some letters are concluded with a courteous sentence to act as a means of signalling the end of the document.

WITH CARE
AIR CARGO HANDLING PLC

Hanger 18R, Manchester Airport, Manchester MR4 6JE
0161 346 98667
email: **withcare@manair.aviation.com**

(A)

(B)

23 March 2006

Mr Peter Phillips
Despatch Department Manager
Mercury Components plc   (C)
Unit 7
Coniston Industrial Park
BARNSLEY
South Yorkshire
SO13 6BN

Dear Mr Phillips   (D)

AIR FREIGHT TO CHICAGO 4 April 2006   (E)

Thank you for your company's recent request to quote for transporting a packing crate to   (F)
Chicago.

As you know, our Mike Richards came to your organisation yesterday to examine the crate, take its
measurements and establish its weight. As a result of his visit we are pleased to be able to quote the sum of
£568.90 + VAT. Our formal quotation is enclosed with this letter.

For this sum we will:

-    collect the crate on 2 April before 12 noon

-    transport it to our depot at Manchester Airport   (G)

-    ensure the paperwork for its journey is in order

-    obtain UK Customs clearance for the crate

-    put it on flight WC457 departing at 15:20 hours on 4 April, for Chicago O'Hare Airport

-    upon arrival, arrange for our American handlers to unload the crate and obtain US Customs
     clearance

-    store safely in the depot until your US client collects the crate.

We trust this quotation is acceptable and look forward to assisting you on this occasion. We would need
confirmation of your wish to employ our services no later than 28 March.

If you wish to discuss this matter further, please do not hesitate to contact me.   (H)

My direct line number is 0161 346 2323.

Yours sincerely

(I)

Paul Falcon
Procurement Manager
Enc

(J)

Key to parts of a business letter

(A) The **letter heading** of the company including a company logo.

(B) **Date** expressed as dd/mm/yyyy.

(C) **Name**, **title** and **company name** and **address** of the person and company receiving the letter.

(D) **Salutation** – Dear Mr Phillips because the letter is addressed to him in the name and address line.

(E) **Heading**: indicating what the letter is about.

(F) **Introductory paragraph**.

(G) **Middle paragraphs** providing details.

(H) **Closing paragraphs** providing an action statement and a courteous close.

(I) **Complimentary close**: Yours sincerely because the recipient's name is used in the salutation. The writer's name and title, leaving space for his signature!

(J) **Enc** indicating there is an enclosure.

Useful phrases for business letters

Thank you for your letter dated ....

As you may know ....

I wish to inform you that ....

I was pleased to hear that ....

I wish to enquire about ....

I should like to place an order for ....

I look forward to hearing from you in the near future.

I should be grateful if you would kindly send me ....

Following our recent telephone conversation, I wish to ....

Please do not hesitate to let me know if I can do anything further to help ....

USING THE TELEPHONE AND MAKING TELEPHONE CALLS

Before you place a call

✓ Think about what you wish to say and how you will say it. Courtesy is expected when using the telephone just as if you are talking in person.

✓ Make a list of what you need to say and the information you need to give and/or receive **before placing the call**. **BE PREPARED**.

✓ Dialling too quickly may be the cause of dialling a wrong number, never just hang up. Apologise and let the person who answered the telephone know you have dialled the incorrect number.

How to speak on the telephone

✓ When speaking, think of the way you sound. On the telephone sounds and moods are magnified. **Talk with a smile in your voice**. The person on the other end of the telephone cannot see your facial expressions and your tone of voice will need to express politeness, enthusiasm and efficiency.

✓ Make sure you say your words clearly and precisely. It is embarrassing, and time-wasting, to be asked to repeat what you are saying. Names and addresses are particularly difficult, so say yours slowly, spelling any unusual words.

Making telephone calls

✓ It is polite, and necessary, to identify yourself. If you are calling from a company, then you would need to identify your company, your name, and perhaps your department, before going on to say why you are calling. For instance:

Good morning, this is Blackwood and Company of York. Janet speaking from the Purchasing Department. I am ringing to place an order...... I wish to speak to ....

How to answer a ringing telephone

✓ The proper way to answer the telephone is give a greeting – **hello; good afternoon** – followed by identifying your telephone number if it is your home, or your name and your company. **Never** answer with just "hello" or "yes". Hello is useless because it does not tell the caller anything, and "yes" is curt and impolite, and again it does not tell the caller anything – except perhaps that you are in a bad mood and cannot be bothered.

Good manners on the telephone

✓ Answer a ringing telephone promptly.

✓ If you dial a number that is wrong, apologise promptly and hang up.

✓ Calling a business at or very near closing time is thoughtless and not likely to result in a successful call.

✓ Introduce yourself when placing a call.

✓ Answer a phone by identifying yourself, your company and/or your department.

✓ When speaking to anyone who is working and for whom time is important, make your call informative and short – plan ahead.

✓ It is polite to let the person who **made** the call **end** the call.

WRITING AND SETTING OUT PERSONAL LETTERS

A personal letter is a letter written from someone's home address to either:

✓ a company – for instance to accompany a job application, or to complain about something; or

✓ a friend – for instance to invite a friend to stay with you.

Some points to remember about letter writing

✓ **Firstly**: the date.

Put the date the letter is written. This date should be shown as:

dd/mm/yyyy

that is: 14th June 2005. Do not mix this order.

✓ **Secondly: the name and address to where the letter is being sent.**

Remember to write to a **person** if you can;

that is: Mr Jaz Allahan.

If you don't know the name of the person, address the letter to a job title;

that is: The Marketing Manager.

> **Remember**
>
> Don't just write Allahan and Corby Ltd. A COMPANY cannot open a letter, but a PERSON can!

If it is an informal letter to a friend, it is acceptable to omit the name and address.

✓ **Thirdly: who are you writing to?**

When you write "Dear ...." it is called the **salutation**.

When you write "Yours ...." it is called the **complimentary close**.

The salutation and complimentary close must match.

That is: Dear Mr Jones = Yours sincerely

Dear Sirs = Yours faithfully

When you use a person's name, be sincere!

> **Note**
>
> Only the word *Yours* has a capital letter at the beginning.

In an informal letter to a friend you can write "Dear Patrick".

If it is an informal letter to a friend you just need to write "Best wishes" or "Kind regards" and sign your first name.

✓ **Fourthly: sign the letter.**

A letter from you needs to be signed. After the complimentary close, leave yourself space for a signature, then print your name. This is important because your signature may not be readable and the person who receives the letter will not know your name.

Examples of address, salutation and complimentary close

Name and address:	Mr P Marks Sunningbrow Golf Course Sunningbrow Hill Aberdeen AB7 3NH
Salutation:	Dear Mr Marks
	Never write Dear Mr P Marks – just Dear Mr Marks. Think of how you would address him if meeting him. You would say "Mr Marks", so write it as you would say it.
Complimentary close:	Yours sincerely
	You have used his name, so be SINCERE!

Name and address:	The Sales Manager McKie and Aston plc 8 School Fields York YO14 5ND
Salutation:	Dear Sir or Madam
	because you have not used a name
Complimentary close:	Yours faithfully
	You have not used a name, so how can you be SINCERE!

Name and address:	Mrs K Trent Office Manager T&N Agency Villamoura Road Bexhill on Sea Sussex SX5 7BQ **This time you have used a name and a job title.**
Salutation:	Dear Mrs Trent
	because you have addressed the letter to her
Complimentary close:	Yours sincerely
	You have used her name, so be SINCERE!

A personal letter written to a company

The following is an example of a personal letter written to a company:

Writer's home address, or return address. Don't put your name here.

6 Telford Drive
Hightown
Wiltshire
HT4 7VV

The address and telephone number/email address can be:

- in the centre;
- at the right hand side;
- at the left hand side; or
- a combination as seen in this example.

Telephone: 01652 974356
Email: trevor@communication.co.uk

Always begin with the date (dd/mm/yyyy).

12 April 2005

The Secretary
Hightown Drama Society
The Strand Theatre
Hightown
Wiltshire
HT2 3GG

Dear Sir or Madam

I wish to enquire if your Society has any vacancies for someone who is a keen amateur dramatist?

In June I will be completing a two-year drama course at Hightown Community College and, before I start University in October, I would be keen to gain experience in a theatre. This last year at College I have particularly enjoyed working behind the scenes and would appreciate any experience of this type, if available.

I look forward to perhaps hearing from you in due course and thank you for considering my request.

Yours faithfully

Leave yourself space for a signature.

Trevor Moore

The letter is addressed to someone's title because the name of the recipient is unknown.
In this way the **salutation** is Dear Sir or Madam
The **complimentary close** is Yours faithfully

**Note: It is usual for a female to write her title: Mrs Tina Moore or Tina Moore (Mrs)
A MAN NEVER CALLS HIMSELF MR. So if you receive a letter from 'T Moore', you can assume it is from a man!**

1 Communication: Reference Sheets

An informal personal letter to a friend

The following is an example of an informal personal letter written to a friend:

Writer's home address, or return address. Don't put your name here.

The address and telephone number/email address can be:

- in the centre;
- at the right hand side;
- at the left hand side.

Middlebank Farm
Trum Gelli
Gynedd
GY11 15MK
Wales

07751345 678
pigriffiths22@farmstead.co.uk

9 June 2006

Dear Molly

I received your letter just today, even though you posted it first-class on 2nd June!

It was great to hear all your news. I imagine you are quite excited about being chief bridesmaid for Julia. Fortunately your sister has consulted you about the colour and style of dress and, I must say from your description, it sounds lovely – not too embarrassingly fussy!

Once the wedding is over you must come and stay for a few days before we fly to Mauritius in August. Just let me know and we'll arrange to meet you at the station.

Mother and father say to send their best wishes to Julia for 2nd July, and of course they wish her and Graeme a happy married life.

I've lots to tell you but it can wait until I see you in August. I don't suppose you have thought of what you are going to take to Mauritius, whereas I have metaphorically packed and unpacked my case dozens of times!

See you soon.

Best wishes

Pam

Leave yourself space for a signature.

WRITING REPORTS

In common with any other business document, a report needs to be planned and, before beginning, you must consider the following:

✓ A report will usually be requested by people who need the information for a specific purpose.

✓ A report differs from an essay in that it is designed to provide information that will be acted on, rather than to be read by people interested in the ideas for their own sake. Because of this, it has a different structure and layout.

✓ **Do not write in the first person**.

✓ **Use the past tense to describe your findings**.

"It was found that......... etc., etc". **It** rather than **I** and the past tense of the verb to find, i.e. **found**.

Points to consider before beginning your report

For whom am I writing the report?

✓ A named individual, or a group of people?

✓ Person(s) who have no knowledge of the subject matter?

✓ What do the readers need to know?

✓ What do the readers already know?

What is my objective?

✓ To inform the readers?

✓ To explain ideas?

✓ To persuade?

✓ To consult?

✓ To transmit ideas or information, facts or findings?

✓ To make recommendations about ways of doing things, making improvements or changes?

What is the context?

✓ Urgent/important?

✓ Routine or "one-off"?

✓ Stand alone, or linked with a presentation?

✓ Sensitive?

What source material?

✓ Is it readily available?

✓ Do I need to do any research?

If you skip the planning stage, poor preparation invariably causes time-consuming problems at a later stage.

Structuring a report

A report is used for reference and is often quite a lengthy document. It has to be clearly structured for you and your readers to quickly find the information it contains.

Parts of a report

The nature of the report will vary from routine reports to complex, non-routine reports. The layout will vary too, yet all reports should have the following common features:

Cover sheet

This should contain the following:

- full title of the report;
- your name;
- the name of the person(s) for whom the report is intended;
- the date.

Terms of reference

This refers to:

- the main subject of the report;
- the scope and purpose of the report;
- the audience who will read it.

The terms of reference should tell you:

- what the report is going to discuss;
- why it is being produced;
- who will read it.

You need to know this information before you begin the process of producing the report:

what? + why? + who? = terms of reference

Title

When you have established the Terms of Reference you can consider the Title.

For any title to be of value it must:

- reflect the terms of reference;
- be precise and refer directly to the subject of the report.

Main sections/findings

The report will need to be divided into various logical sections and sub-sections. Make full use of **paragraph headings** and **paragraph numbering** including **bullet points**.

This is the section in which you:

- ✓ state what you found out;
- ✓ clearly present your results, making use of paragraphs, paragraph headings, bullet points, etc.;
- ✓ list the essential data. You may want to use tables, graphs and figures.

Use a consistent system of display throughout. Numbered paragraphs might be 1, 2, 3, or 1.1, 1.2, 1.3, etc.

Using effective "signposting" in this way will help the reader pick out elements of the report and will ensure the whole document is easy to follow.

Conclusions

Remember, the purpose of a report is to provide findings and draw conclusions from those findings. This section is vital in a report and allows the key arguments and findings of the report to be drawn together and put into context.

Conclusions need to:

- ✓ refer to the purpose of the report;
- ✓ state the main points arising in the report;
- ✓ be brief and concise.

Recommendations

Any recommendations you make must be presented clearly and follow logically from the conclusions.

This section might, for example, suggest a preferred option from several that were under consideration, make new proposals or recommend further research or investigation.

Bibliography

List all the sources that you have referred to in the main sections of the report. (*See page 26 on Writing a bibliography*.)

A sample title page of a report

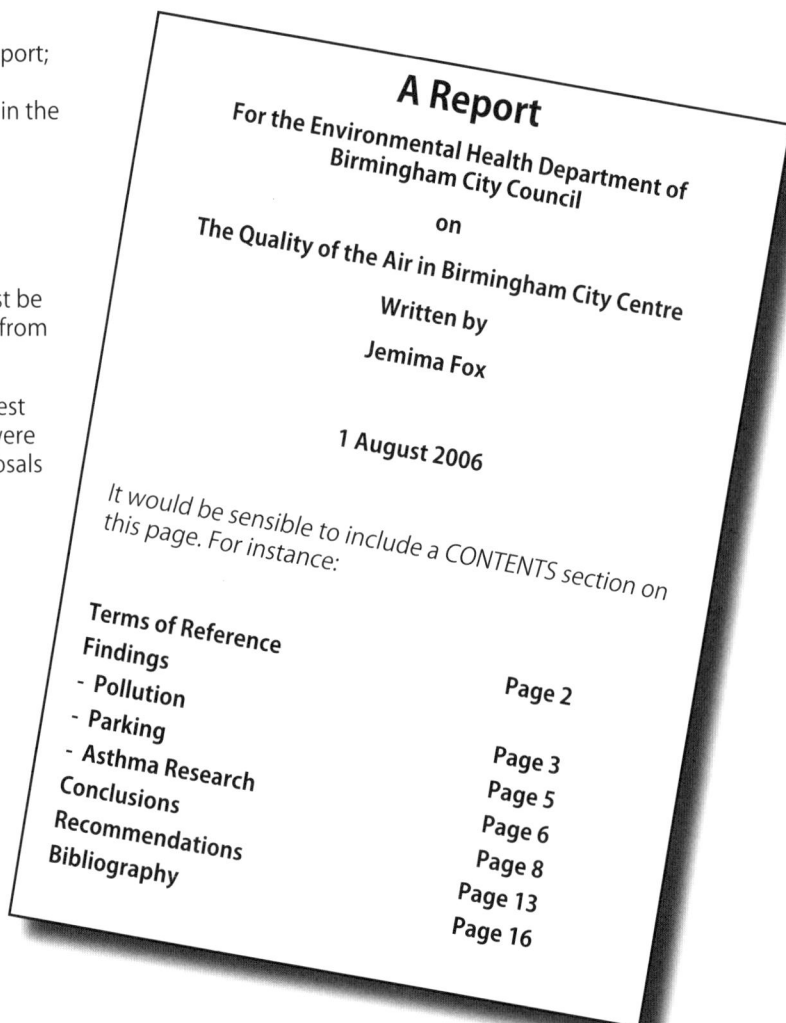

A Report

For the Environmental Health Department of Birmingham City Council

on

The Quality of the Air in Birmingham City Centre

Written by

Jemima Fox

1 August 2006

It would be sensible to include a CONTENTS section on this page. For instance:

The report's content

Terms of reference

Report on "The Quality of the Air in Birmingham City Centre" for the Environmental Health Department of the City Council, in order that it can consider possible traffic limiting proposals in the City Centre during peak weekday periods.

Title

The Quality of the Air in Birmingham City Centre

Findings

This report was requested by   (*person(s)*) in order that   (*what the persons are going to do with the report findings*).

In order to produce this report it was necessary to conduct research into the following topics:

....

....

....

and the sources consulted are listed separately in the Bibliography on page X.

1   Pollution.

       Make an opening statement about the topic and its relevance to the report then separate into sub-headings, which may, or may not, be numbered.

1.1   UK city average pollution figures.

1.2   City pollution figures related to increase in the number of Asthma sufferers.

1.3   etc., etc., etc.

2   Parking.

Conclusions

The purpose of this Report was to .... As can be understood from the information contained in "Findings", there is overwhelming evidence that the City of Birmingham's City Centre has the second highest pollution figures in the UK................ etc., etc., etc.

Go on to summarise your findings.

Recommendations

The Environmental Health Committee will consider the content of this report when deciding whether to limit traffic in the City Centre during peak periods of the week.

In view of the evidence, the recommendations to the Committee are as follows:

Describe the recommendations, making use of paragraphs, paragraph headings, bullet points, underlined headings, emboldened text, etc.

A report is a business document so you must sign it at the end.

Bibliography

List the documents you consulted in your research here.

Writing a bibliography

A **bibliography** is a list of sources you have consulted and used information from, for such documents as a report, essay or presentation.

The bibliography is usually produced as an **Appendix** and the information is displayed in categories (books, reference books, newspaper articles, web sites). Each **category heading** appears in alphabetical order and under each **category heading**, the sources are arranged alphabetically. (*See Example 1 below.*)

Example 1

Bibliography

Books

Anderson, M, 2005, *The history of needlework*, Swan & Signet Books, Preston.

Connaught, J P, 2001, *British needlework explained*, Mercantile Books, Edinburgh.

Newspaper Articles

Silvester, C, 2003, 'UK needlework trends since 1851', *The Nottingham Gazette*, 7 April 2003, p15.

Uxbridge, K, 2004, 'Needles and Threads', *The Cornish Messenger*, 20 September 2004, pp49 and 50.

Reference Books

The Encyclopedia of European Embroidery, 2002, 3rd edition, Stitch-in-Time Publishing, Worcester.

Threads Galore, 2005, 1st edition, Sharp Publishing, Norwich.

Web Sites

Ruby, J T 2006, The UK Needlework Association, Nottingham, viewed 24 August 2006 www.uk-nan.org.uk.

Westward, H and Prentice, B 2005, Simply Needlework, Colchester, viewed 25 August 2006

www.simplyneedlework.com.

There is a recommended **list** of information that should be included in each category, and a recommended **order** in which this information should be presented. Some of this information is separated by a comma and some appears in *italics*.

Books	Newspaper articles
Author's name, comma, initial(s), comma	Author's name, comma
Date of publication, comma	Initial(s), comma
Title in italics, comma	Date, comma
Publisher, comma	Title of article in **single quotations (')**, comma
City/town of publication (not country), full stop.	*Title of newspaper* in italics, comma
	Date of publication, comma
	Page number(s) of the article, full stop.
Reference books	**Web sites**
Title of book in italics, comma	Author (person or organisation)
Year, comma	Date (site created or updated), comma
Edition, comma	Name of publisher, comma
Publisher, comma	City of the publisher, comma
City of publication full stop.	Date you viewed the site
	URL, full stop.

GIVING A TALK OR PRESENTATION

The ability to make an oral presentation is an important skill. In today's workplace employees will be required to address colleagues, or external groups, from time to time, and it is also increasingly common for some job interviews to include an oral presentation.

The five main stages in preparing for a talk or presentation

STAGE 1
Address and answer the main questions related to the topic, i.e. **be clear about the purpose of the talk or presentation**.

STAGE 2
Be clear about your brief, i.e. **work out exactly who is your audience, what they know about the topic already (if anything), and what you intend them to know afterwards**.

STAGE 3
Decide what to include, and what **not** to include.

STAGE 4
Decide how to organise the material.

STAGE 5
Make the structure and sequence **logical**.

> "Your brain starts working from the moment you are born and never stops, until you stand up to speak in public." *Sir George Jessel*

> **THOROUGH PREPARATION IS VITAL TO SUCCESS**

Delivering your presentation

- It is preferable to stand when you are speaking unless it is an informal presentation to a small group.

- Ask your audience if they can see and hear you and any material you may be using (overhead projector slides (OHPs), whiteboards, flipcharts, etc.)

- Try to use your OHPs or software slides as a prompt so that you do not have to use notes.

- If you do not feel confident enough to do this, then notes on cards are easy to hold. If you use word-processed notes, use a typeface that you can read easily and a type size of 14 point or more and space them out so that they are easy to see at a glance. **Highlighting** key words helps.

- Do not use technical jargon without being prepared to explain it to your audience. Do not assume your audience is as familiar with your topic as you are.

- Do not speak too fast. Vary your pace and, if your voice is normally quiet, then speak a little louder than usual.

- Keep eye contact with your audience. Include the people on the edges of the group. Eye contact keeps your audience engaged in your presentation. Do not address the projector screen!

- Do not fiddle with pens, pointers, etc. when speaking. Do not jangle keys or cash in your pocket. Try to keep still.

- If you are one member of a group making a presentation decide how you are going to divide the presentation up? Who is going to do what?

Images

Amongst other things, images serve to:

- ✅ add interest to what you are saying;
- ✅ focus the audience's attention;
- ✅ clarify facts;
- ✅ help the audience remember what you say.

Never read it out!

Nervous speakers often make the mistake of reading from their script. This tactic results in a head down, zero eye-contact, monotonous lecture, not an interesting talk with the speaker engaging the audience's attention and interest.

Of course, you know your topic well, having prepared it well, so now prepare a series of **cues** in the form of notes on cards to prompt you about your topic. Your talk will be based on these cues, not on a script!

Creating an opener

At the beginning tell the audience the context of your talk and a brief outline of what you are going to cover.

How do you first secure the attention of the audience?

They will want to know two things:

- ✅ that your message is relevant and interesting to them; and
- ✅ that you have the presence/credibility necessary to deliver it.

What openers can you use?

Ask a relevant question "Let me begin by asking you a question ...."

Quote a statistic "Did you know that X% of people in the UK ...."

Use a relevant quotation

Follow these up with a brief explanation of **who you are** and **what you are going to talk about**, emphasising what members of the audience will gain by listening.

Closing with a flourish

The close is the equivalent to the conclusion section of a report or a letter – where you draw arguments and facts together.

Remind everyone of the objectives of the talk/presentation, and summarise your key points.

"So before I finish, I'd like to summarise the points we've covered this afternoon ...."

It's question time! – Handling questions

- ✅ Allow time at the end of your presentation for questions.
- ✅ Tell your audience at the beginning when you want to receive questions, to avoid interruptions. For short, time-limited presentations it is best to leave questions to the end. Otherwise you may not have time to finish your presentation.

TASK DESCRIPTION GRID

Number and title	Page	Activities	Refer to reference sheet(s) on page(s)
1 A1 Motors	30	Writing a memo.	6 – 7
2 Telephone Messages	32	Completing telephone message forms.	8
3 Finch and Rook	33	Completing an accident report form. Writing a memo.	6 – 7
4 Ladybird Nursery	34	Writing a memo.	6 – 7
5 The Contented Plaice	35	Taking part in a discussion with a working partner. Designing a poster and including images.	10 – 11
6 Blooming Plants	36	Working with a partner. Designing an advertisement. Writing a business letter.	12 – 13 14 – 16
7 Home Comforts	38	Rewriting a business letter.	14 – 16
8 Home Comforts	40	Writing two business letters.	14 – 16
9 Swan Theatre	43	Writing a business letter.	14 – 16
10 Railway Timetables	45	Researching and writing documents.	
11 Planning a Journey (Derby)	46	Researching and making a telephone call. Writing a personal letter.	17 20
12 Planning a Journey (Lincoln)	47	Researching and writing a personal letter.	18 – 21
13 Pickton Lift Company	48	Researching and writing memos.	6 – 7
14 Hop, Skip and Jump	49	Completing forms, working with a partner to make telephone calls. Completing a telephone message form.	17 8
15 Value for Money	52	Completing a form.	
16 Pets Safe at Home	53	Designing a newspaper advertisement. Writing a business letter.	12 – 13 14 – 16
17 Currently the Best	54	Researching and writing a personal letter.	16, 18 –20
18 Recycling	56	Researching and writing a report. Giving a short talk and taking part in a discussion.	10 – 11, 22 – 26 27 – 28

TASK 1: A1 MOTORS

Student Information	REMEMBER:
In this task, you will write a memo, using information contained in **Appendices 1** and **2**. Ask your tutor to provide you with the blank **A1 Motors memo sheet**.	A memorandum is a brief, internal written communication. It can be a formal document (*see page 6* for an example). It can be an informal document (*see page 7* for an example). It is signed by the writer.

Writing a memo

Scenario

You work in the Customer Liaison Department of a garage called **A1 Motors**. Today is Friday and you have to inform the Service Department Manager of the number of cars booked in, through your Department, for servicing on Monday to Wednesday of next week.

Activities

Appendix 1 gives details of the cars booked in to the Service Department on Monday to Wednesday inclusive, next week. **Appendix 2** gives details of the registration numbers linked to customers' names.

1   Write the memo from yourself, as the Administration Assistant of the Customer Liaison Department.

Send it to Ben Trent, Service Department Manager.

2   Use the heading: **Cars booked in for servicing Monday to Wednesday** (dates of next week please).

You will need to supply the **car registration**, its **make** and **model**, and the **owner's name**. A table might be the best form of displaying this information clearly. Think of the best way of displaying the cars' details on each day of the week.

Appendix 1

```
Cars booked in . . . .

    Monday          X678 TAN  (two-day service and repair)
                    FR52 OPL
                    C48 TOH

    Tuesday         SM51 GGR
                    NU03 FPQ
                    S1 MON

    Wednesday       KN02 YKZ
                    X231 NOP
                    NW51 PLC
                    NL03 YSY
```

Customer Name	Car Make and Model	Vehicle Registration
Pettigrew Charles, Mr	Vauxhall Cavalier	SM51 BVS
Collin Simon, Mr	Fiat Uno	S1 MON
Sturt Adrian, Mr	Vauxhall Vectra	KN02 YKZ
Jenkins Marsha, Mrs	Toyota LandCruiser	NL03 YSY
Clarke Bryan, Mr	Rover 45	NU03 FPQ
Stephenson Craig, Mr	Mini Cooper	SM51 GGR
Stevenson Colin, Mr	Mini Cooper	X231 NOP
Shute Martin, Mr	Range Rover	RR53 RVR
Harton Carol, Miss	BMW 5 series	FR52 OPL
Callisto Henry, Mr	Rover 25	C48 TOH
Jacobs Pauline, Mrs	Renault Clio	X678 TAN
Brentwood Alison, Miss	Rover 75	NW51 PLC

TASK 2: TELEPHONE MESSAGES

<table>
<tr>
<td>

Student Information

In this task, you will complete two telephone message sheets.

Ask your tutor to provide you with the blank **telephone message sheets**.

</td>
<td>

REMEMBER:

Telephone messages are brief.

See page 8 for details of how to take messages.

Telephone messages contain only relevant information.

The information you include must be accurate.

Study the sheets carefully to make sure you complete all the important parts.

</td>
</tr>
</table>

Telephone messages

Scenario

You work in a local firm and today take two telephone messages that have to be passed to colleagues.

Activities

1   In handwriting, taking special care of spelling and punctuation, complete the two sheets with the details of the calls. These can be found in **Appendix 1**.

In each case you enter your own name as having taken the message and the time is 10:50.

> **Note**
>
> A completed telephone message sheet contains only relevant information so you need to omit anything which is not necessary to the meaning of the message. Be careful to be accurate in what you write.

Appendix 1

Message 1 – for John Stone, European Director

Daniel Patterson of JK&M Ltd, Denmark, wants to confirm he will be accompanying you on your trip to Lisbon next week.

However, he cannot leave on Monday, as you are doing, because of another appointment which cannot be cancelled. He wants you to know he is leaving on Tuesday on the 08:45 flight from Copenhagen, arriving in Lisbon at 12:20.

If you really want someone from his organisation to be with you on Monday, and he is not sure if you do or not, then he can suggest Erica Lindstrom.

Ring to confirm what you want him to do. He will be out of the office until 4pm our time. You have his number but before 4pm you can reach him by email on dpatterson@jkm.marathon.co.dk.

Message 2 – for Seth Stavely, Sales Manager

Mary Benson of Finch and Rook cannot attend tomorrow's meeting at 10:15. She is sending Peter Farraday in her place. You are requested to give Mary a ring to confirm you agree with the change. Her number is 01904 357 887 (extension 223). She will be in the office until 5pm.

If the call is to be after 6pm then ring her at home on 01904 673 456.

TASK 3: FINCH AND ROOK

Student Information	REMEMBER:
In this task, you will complete an accident report form and write a memo. Ask your tutor for a blank **Finch and Rook Accident report form** and **memo sheet**.	The information you complete must be accurate and relevant. Study the sheet carefully to make sure you complete all the relevant parts. Your memo should be brief and you must sign it. *See pages 6 and 7 on how to write memos.*

Accident report form and writing a memo

Scenario

You work in the offices of **Finch and Rook**, in the Publicity Department and today have to complete an Accident Form to report an accident you saw take place in the restaurant at work.

The accident happened today.

Activities

1　　　Complete the **accident report form** so that it reflects relevant information shown below:

> ### Accident information
>
> - At 12:10 today you were in the staff restaurant queuing at the counter for food.
> - Michael Winchester was in front of you when he dropped his tray of tea and orange juice on the floor.
> - Next to Michael was Sophia Baxter and as she stepped forward, she slipped on the wet floor.
> - No one had had time to clean the area because immediately after Michael dropped the tray Sophia stepped forward.
> - As Sophia fell her leg twisted under her and she was in obvious pain. She was lying on the floor.
> - The First Aider – Tom Prentice – was called and he suspected a broken ankle and asked for an ambulance to be called. You called for the ambulance and there was a wait of 20 minutes and Sophia was taken to the Wellbeing University Hospital in the town centre at 13:25.
> - She was seen in casualty and diagnosed as having a sprained ankle. Tom was incorrect in his diagnosis of a broken ankle.
> - She was bandaged and sent home at 15:30.

2　　　Having successfully completed the form, send a **memo** to Andrew Goldton, the Health and Safety Officer, enclosing your form. Use a suitable heading and content in the memo.

> ### Remember
> Your form must be completed neatly and contain accurate and relevant information. Correct spelling and punctuation are important too.

TASK 4: LADYBIRD NURSERY

Student Information

In this task, you will write a memo containing information you have extracted from **Appendix 1**.

Ask your tutor for a blank **Ladybird Nursery memo sheet**.

REMEMBER:

Display the information in a way that is easy to interpret.

Make sure the information you give is accurate.

You must sign the memo.

See pages 6 and 7 on how to write memos.

Interpreting graphical information

Scenario

You work in the office of the **Ladybird Nursery**, a children's nursery situated in your town/village . The Coordinator of the nursery is Carol Button and she has asked you to prepare information related to next month's bookings in the nursery.

Activities

1   You are to write a memo to the Nursery Coordinator, Pip Layton.

The purpose of the memo will be to give the Coordinator details of any child who is new to the nursery. This information is needed as it allows staff to better prepare for the arrival of a child new to nursery life.

Extract relevant information from **Appendix 1** to include in your memo.

You will need to provide information on the name of the child, which days and session they will attend, and any other information you consider relevant. To present an ordered and structured document, the names should be in alphabetical order.

Remember to give the memo a suitable subject heading.

Appendix 1

BOOKING RECORD FOR THE MONTH OF *NEXT MONTH*

M		T		W		Th		F		Name of Child
am	pm	am	pm	am	pm	am	pm	am	pm	
✓		✓		✓		✓		✓	✓	Christopher Devlin
	✓		✓		✓		✓		✓	Martha Sutton
		✓	✓			✓	✓	✓		David Squires
✓		✓		✓		✓		✓		**Dene Fletcher**
	✓	✓		✓	✓			✓		**Dene Pollock**
✓		✓		✓		✓				Steven Jefferson
	✓	✓		✓			✓	✓		**Carla Hindmarsh**
✓	✓		✓	✓	✓		✓	✓	✓	**Claire Higginson**
	✓		✓		✓		✓		✓	Katy Fletcher
		✓	✓					✓	✓	Thomas Bentley
				✓	✓	✓	✓			**Louis Trent**
	✓			✓		✓		✓		Gail Breeze
✓		✓		✓		✓			✓	**Anne Forbes**
			✓			✓		✓		Victoria Sykes
	✓	✓			✓				✓	Jamie Westerbrook

Shaded rows = new child starting

TASK 5: THE CONTENTED PLAICE

Student Information
In this task, you will work with a partner to plan a poster to advertise a restaurant.

REMEMBER:
Make notes of the discussion you have with your partner.

Use the notes to design a poster.

Design the poster and include relevant image(s).

See pages 10 and 11 on Using Images in Communication.

Ask your tutor if you can have a photocopy of the poster so that you each have a copy.

Working with a partner and designing a poster

Scenario
You are working in a restaurant in Your Town called The Contented Plaice. It serves mainly fish and chips, but has recently begun to offer a variety of snack meals and it is this change that the restaurant wants to advertise.

Today you are to work on an advertisement poster.

Activities

With a partner discuss what will be required to successfully complete the tasks. Then produce the poster together.

Make notes of the things you discuss and keep those notes as evidence. The notes must be organised in such a way that you can transfer topics you have discussed onto your poster.

1   Design an A4 poster (one side only) to advertise the new speciality snacks. Once complete, the poster will go to the local printing company that will print lots of copies for distribution around Your Town.

2   These dishes are:

   - jacket potatoes;

   - toasted sandwiches with a variety of hot fillings;

   - hot dogs; and

   - hamburgers.

3   Include relevant images in your poster and any other information you think necessary.

4   The restaurant is open from 11am until 4pm Monday – Friday and 10am to 6pm on Saturdays.

TASK 6: BLOOMING PLANTS

Student Information	REMEMBER:
In this task, you will work with a partner to design a newspaper advertisement. Individually write a business letter to the local newspaper asking for the advertisement to be inserted. Ask your tutor for a blank **Blooming Plants letter heading** and a blank copy of **Appendix 2 Pro Forma**.	Make notes of the discussion you have with your partner on the form provided. Use the notes to design the advertisement (*see pages 12 and 13*). A business letter is a formal document (*see pages 14 – 16*). Make sure it contains the following: ● the date it was written; ● the name and address of the recipient; ● a salutation; ● a complimentary close that matches the salutation; ● the name of the person who wrote the letter.

Working with a partner, writing an advertisement and business letter

Scenario

You work for **Blooming Plants**, a garden centre. On the 21st of next month the centre is moving to larger premises. It is your job to deal with two aspects of this move.

Activities

With a partner

Before you can complete this task each of you will have to show you are clear about the objectives of the task and the information you will need to successfully complete the task. For this purpose you will need to complete **Appendix 2 Pro Forma**, which will help with the planning process.

1   Study the information contained in **Appendix 1**.

2   Design a newspaper advertisement to advertise the forthcoming move. It will go in the local paper – *The Bugle*. You should make use of appropriate images.

       Your advertisement should be aimed at existing and potential customers.

       It should give the date of the move; the last date of trading in its current premises; the opening times in the new premises; the advantages to the customer. Include other appropriate information.

3   To encourage new customers, your firm is going to have a special offer for its first day in the new premises. Decide, together, what that special offer is to be and include details in your advertisement.

       Note: You will each need a copy of the advertisement for your individual evidence.

Individually

4   Use the letter heading to write to The Business Advertising Manager, The Bugle, Lord Fullerton's Walk, Your Town YO1 7VQ.4

       Ask for the advertisement to be placed in the paper on the 1st, 8th, 10th, 15th and 16th of next month.

       For the first three insertions the advertisement should be a quarter-page spread in colour, the remaining insertions you wish to be a colour half-page spread.

       Ask for the bill to be sent to the centre's Finance Manager – Bill Cash.

BLOOMING PLANTS

Springbank Lane, Your Town, YO19 5PW

01863- 567 248
blooming@lexodus.co.uk

STAFF NOTICE

As you are aware, with effect from 21st of next month, we will be moving to larger premises and trading from **Vicarage View, Denton, Your Town YO16 4MS**   Our telephone number will be unchanged.

We will cease trading at 7:00pm on 18th and open at 8:30am on the 21st in Vicarage View.

The new location has the advantage of being much larger than our existing site. It is almost three times the size and we have been able to incorporate the following, new, amenities:

* indoor plant and equipment centre **with specialist staff on duty every hour of opening**

* outdoor plant and accessories centre **with specialist staff on duty every hour of opening**

* water garden centre **with specialist staff available throughout opening hours**

* Japanese garden centre

* greenhouse centre **with specialist staff to design new greenhouses**

* garden design centre, open Tuesdays and Saturdays.

New Opening Hours

beginning of April to end of September		beginning of October to end of March	
Monday	09:00 – 18:00	Monday to Friday	09:30 – 17:00
Tuesday to Friday	08:30 – 18:30	Saturday	09:00 – 17:30
Saturday	08:30– 19:00	Sunday	10:00 – 14:00
Sunday	10:00 – 15:00		

We are sure the relocation to larger premises will enable us to provide even better services to our existing customers and we hope to attract new customers. Thank you for your co-operation during what will be a busy time.

Holly Bush
Centre Manager

TASK 7: HOME COMFORTS

Student Information

In this task, you will correct a draft business letter that contains spelling errors and is laid out incorrectly.

Ask your tutor for a blank **Home Comforts letter heading**.

REMEMBER:

The letter is known as a **circular letter** because the same letter is being sent to lots of customers.

Make sure it contains the following:
- the date it was written;
- a space where the name and address of the recipient can be added at a later date;
- a salutation;
- a complimentary close that matches the salutation;
- the name of the person who wrote the letter;
- the title of the person who wrote the letter.

See pages 14 – 16 for an example of a business letter.

Writing a business letter

Scenario

You have recently begun working for a soft furnishings company called **Home Comforts** and the Sales Manager, Neil Garner, has handed you a letter he has drafted to his prospective customers.

It is your job today, as his assistant, to ensure a correctly-displayed and correctly-spelt and punctuated letter is ready to leave the firm.

Activities

1   His draft can be seen in **Appendix 1**. You know there are spelling, punctuation and display errors.

2   Prepare a correct version of the letter on the blank letter heading included.

> **Remember:**
>
> Your letter must be laid out correctly and your meaning must be clear.
>
> Correct spelling and punctuation are important too.

HOME COMFORTS

Furnishings of Distinction

119 Mandolin Square   LINCOLN   Lincolnshire   LN1 3BV

Phone:015783644562 **Fax:**015783644652 **email:**HomeCom@Krypton.co.uk

Dear Sir/Madame

I want to take this oportunity to introduce myself and my companie. I have been the Manager of this company since 1987 and my employees and I are all dedicated to providing our customer's with an excellant service to match our top quality goods.

The purpose of this letter is to let you no that for the whole of next month we are offering a outstanding 35% discount of all Grade A carpets and all three-piece suits in the Excelsior range. Furthermore, their will be a 20% discount off all beds and a 10% discount off kitchen ware.

I hope you will visit the store because I am sure you will be delighted with the quality of our goods and the excellence of our service. Alot of our customers return to us year after year and to us, that says our policy of customer care is sucessful.

We all look forward to seeing you next month.

Yours sincerely

Neil Garner   Sales Manager

TASK 8: HOME COMFORTS

Student Information

In this task, you will write two business letters replying to complaints from customers.

Ask your tutor for two blank **Home Comforts letter headings**.

REMEMBER:

Make sure each letter contains the following:
- the date it was written;
- the name and address of the recipient;
- a salutation;
- a complimentary close that matches the salutation;
- the name of the person who wrote the letter;
- the title of the person who wrote the letter.

Include only relevant information and make sure the tone is an appropriate one to pacify a complaining customer.

See pages 14 – 16 for an example of a business letter and page 16 for useful phrases to include in a business letter.

Writing business letters

Scenario

You are the assistant to Peter Burton, the Manager of **Home Comforts Department Store**.

Today Peter has received the attached letters from two disgruntled customers and asked you to reply on his behalf.

Activities

1   Letter from Mr Russon

Apologise. Say you have followed the matter up today and the cheque is in the Accounts Department awaiting the signature of the Head of that Department. You have been assured by him – John Springer – that the cheque will be in tonight's first-class post. If Mr Russon has any queries he should ring Mr Springer on his direct line number – 377 83347.

2   Letter from Mr Cross

Apologise. Say you regret the problem, etc. etc. You have ordered a new sofa in Atlantic Blue today. Manufacturers have been asked to expedite the order and have indicated delivery in three weeks from today. You will contact the customer when the sofa is in the store and arrange delivery.

Offer him the opportunity, once he has a matching suite, of coming in to the store to select up to six cushions that will match his blue suite.

24 Salisbury Avenue
IPSWICH
IP6 4CR

(dated yesterday)

Mr P Burton
The Manager
Home Comforts
119 Mandolin Square
LINCOLN   LN1 3BV

Dear Mr Burton

<u>KITCHEN TABLE</u>

You will recall that ten days ago I returned to your store a kitchen table that was delivered with deep cuts in its surface. This was despite having been inspected by myself in the store and deemed to be in "perfect" condition.

It was agreed over the telephone 7 days ago, that once the table was returned I would receive a refund for the full price - £160.

To date, I have not received this money and wish you to reassure me that I will receive payment within the next five working days.

It hardly needs to be said that I am extremely disappointed, not only in the quality of your products but in your inability to refund money which is due. Should this matter not reach an acceptable conclusion within the time stated, I will have no alternative but to contact the Trading Standards Department of our local Council.

Yours sincerely

Charles Russon

C Russon

"Tulip Cottage"
Milton Green
SUFFOLK
MP3 5BQ

01455 367 3888

(DATED TWO DAYS AGO)

Store Manager
Home Comforts
119 Mandolin Square
LINCOLN
Lincolnshire   LN1 3BV

Dear Sir or Madam

Three-piece suite – Style "Madrid"

I placed an order six weeks ago for a three-piece suite of the above-named style in "Atlantic Blue".

Yesterday the suite was delivered. Unfortunately when my wife and I unpacked it we found that the chairs were of the colour ordered but the sofa is pale brown – called by your delivery men "Fern Brown".

I spoke immediately to the Department Manager – Catherine Welsh – who advised me to write to you to request replacement of the sofa.

Mrs Welsh allowed me to keep the wrongly-coloured sofa as we have nowhere else to sit – two chairs being insufficient for my wife, myself and our three children.

I would be grateful if you would confirm that a replacement sofa will be ordered and delivered urgently. I do not particularly wish to have the unrequired item for any longer than is absolutely necessary.

I look forward to receiving your confirmation that the above request will be actioned.

Yours faithfully

I A M Cross

I A M Cross

TASK 9: SWAN THEATRE

Student Information

In this task, you will read a telephone message from a colleague who dealt with a complaining customer.

You will then write a business letter replying to the complaint.

Ask your tutor for a blank **Swan Theatre letter heading**.

REMEMBER:

Make sure the letter contains the following:

- the date it was written;
- the name and address of the recipient;
- a salutation;
- a complimentary close which matches the salutation;
- the name of the person who wrote the letter;
- the title of the person who wrote the letter;
- the Encs mark after the writer's name and title (to show there is something enclosed with the letter).

Include only relevant information and make sure the tone is an appropriate one to pacify a complaining customer.

See pages 14 – 16 for business letter examples and useful phrases.

Writing a business letter

Scenario

You work in the Customer Liaison Department of the **Swan Theatre, Stratford upon Avon, Warwickshire** and today have to deal with the complaint of a dissatisfied customer. You are replying on behalf of the Theatre Manager, Sunita Patel, so make sure the letter is written as if from her.

Activities

1 Use the letter heading to reply to the complaint which is detailed in the Telephone Message Form below:

TELEPHONE MESSAGE

Taken by: Box Office Manger, Petra Swindles   **Date:** (today)   **Time:** 09:40

From: Cynthia Bumper

To: Theatre Manager, Sunita Patel

Re:

Mrs Bumper arrived at the theatre last night to see "Goodwill in Spring". She had two tickets, purchased by telephone, in Box 7 (Seats 3 and 4).

When she got to her box the seats were already occupied. She reported this to the Duty Manager, Christian Salvason, who allocated her seats in Row T, Numbers 11 and 12.

Mrs Bumper is annoyed, particularly as she purchased the tickets, and received them through the post three weeks ago, and because the seats she and her husband had to sit in were some distance from the stage. They cost £5.00 each, whilst her Box tickets cost £11.00 each.

The Duty Manager assured her a refund would be available if she wrote or telephoned. This she is now doing. She is very unhappy.

Bear in mind the following details for your reply:

- your task is to apologise;
- say that the Duty Manager offered you the seats in Row T because the performance was due to start in only ten minutes and he was anxious to get you seated so you could enjoy the show;
- enclose a cheque for £22 – full refund of the 2 x £11 tickets;
- enclose a £10 voucher which can be redeemed against any matinee performance next month;
- the shows next month, from which she can choose, are as follows:

1st – 7th	Much Ado About Nothing
8th – 15th	The Neighbour's Cat
16th – 18th	Friends for a Day
19th – 25th	Trial by Jury
26th – end of the month	The Boyfriend.

Mrs Bumper's address is:

8 Halfpenny Walk, Stratford upon Avon, Warwickshire WA5 3SF

TASK 10: RAILWAY TIMETABLES

Student Information

In this task, you will locate, and use, a railway timetable in order to find a suitable train.

You will also locate a map of the UK, and mark on it a number of towns.

REMEMBER:

Include your research documents with the task you hand to your tutor.

If you cannot do this, take a photocopy of the documents you find, and attach the copy to the task.

Make sure that all the information you give is accurate and relevant and displayed in an appropriate way so it is easy to understand.

Reading railway timetables and marking places on a map

Scenario

You live in (select a town with a railway station) and it is your Aunt's birthday on the 15th of next month. She lives in (select another town with a railway station about 150 miles away from yours) and your family has decided to treat her to dinner and overnight accommodation on the evening of her birthday.

Locate a railway timetable and map of the UK with which to work on this task.

Activities

1   **Decide upon a train**

It is your intention to meet your Aunt around mid-afternoon, have a look around the shops, and perhaps take afternoon tea, before going to the hotel where she will stay the night and at which the celebration dinner will be held.

Select a suitably-timed train. She would like to be able to make a seat reservation and to have a buffet meal on the train.

2   Decide upon a suitable train on which she can return home the day after her birthday.

She will not want to leave too early as she will probably have a late night with her family the evening before.

Select a suitable train for her return journey, taking into account the previously-stated recommendations about her return home, and her wish to have a reserved seat and, possibly, a meal on board.

3   **Annotating a map**

Obtain a map which shows her home town and yours, and mark on it the place(s) at which the train stops on both its journeys.

TASK 11: PLANNING A JOURNEY TO DERBY

Student Information

In this task, you will locate, and use, a map of the UK, and plan a car journey to a number of towns.

You will locate information about hotels in Derby and select an appropriate one.

You will take part in a telephone conversation to the hotel, with your Tutor acting as hotel booking clerk, in order to make a booking.

Your last task will be to write a personal letter to the hotel, confirming your reservation.

REMEMBER:

Include your research documents with the task you hand to your tutor.

If you cannot do this, take a photocopy of the documents you find, and attach the copy to the task.

Make notes of what you will say on the telephone before you place the call. This will help you to get the conversation right.

See page 17 for guidance on Using the Telephone.

Make notes of what is said to you during the conversation because you will have to use some of this in your personal letter.

A personal letter has your address at the top and the rest of the letter is laid out as a business letter. In this case your personal letter is a formal letter. (*See page 20 for an example.*)

Planning a journey by road, booking hotel accommodation by telephone and writing a personal letter to an hotel

Scenario

Next month, you and a friend, who lives in Rotherham, South Yorkshire, are going to spend a few days in Derby. You live in Newcastle upon Tyne. It is fallen to you to arrange the hotel accommodation and plan the car journey from your home, via your friend's and on to Derby.

You must include your source documents.

Activities

1   Using a suitable source – plan the car journey, marking all the major roads upon which you will travel. You will leave your home, drive to Rotherham, pick up your friend and then drive to Derby.

2   Using an appropriate source, find a suitable hotel in Derby.

     You want bed and breakfast for four nights: the first Thursday to Sunday inclusive of next month.

     The accommodation should be close to the city centre and you each have a budget of £50 per night.

3   With your tutor as the Reservations Manager of the hotel of your choice, telephone and make the booking. You will be asked to confirm the booking in writing, and may be asked for other details.

     Make notes because your next task is to confirm the booking.

4   Write to the Reservations Manager, confirm the booking and provide any other information requested in the phone call.

> **Remember you must use appropriate formats for presenting your information, adopting an appropriate style and taking care that your meaning is clear and that spelling, punctuation and grammar are accurate.**

TASK 12: PLANNING A JOURNEY TO LINCOLN

Student Information

In this task, you will locate, and use, a map of the UK, marking towns on the map.

You will locate information about hotels in Manchester and select an appropriate one.

Your last activity will be to write a personal letter to a friend, confirming your forthcoming stay in Manchester.

REMEMBER:

Include your research documents with the task you hand to your tutor.

If you cannot do this, take a photocopy of the documents you find, and attach the copy to the task.

A personal letter has your address at the top and the rest of the letter is laid out as a business letter. (*See pages 218 – 20.*) In this case your personal letter is an informal letter. (*See page 21* for an example.)

Planning journey by road, booking hotel accommodation and writing a personal letter

Scenario

On Friday next week, you and your friend want to travel to Manchester by car. Your friend lives in Lincoln and you plan to drive from your home town of Sheffield to Lincoln to pick him/her up, then drive to Manchester and stay in a city centre hotel for three nights, before reversing the journey.

Activities

1   In order to have some idea of which parts of the country you will be visiting, mark Manchester, Lincoln and your home town of Sheffield on the map.

   You will, of course, need to consult a reliable reference source in order to do this task accurately.

2   Use a UK Hotel Guide to find a suitable hotel in Manchester in which you can stay for bed and breakfast. You wish to be in the city centre and each has a budget of £60 per night.

   Make a note of the name, address and telephone number and details of any facilities it offers, e.g. swimming pool, leisure complex, etc.

3   Write a personal letter to your friend.

   Advise him/her of the hotel you have chosen and tell him/her the day and date you will collect him/her from their home.

   Add any other information you consider relevant.

TASK 13: PICKTON LIFT COMPANY

Student Information

In this task, you will locate, and use, temperature details about two cities abroad.

You will use the information you find to write two memos to members of staff.

Ask your tutor for two blank **Pickton Lift Company Memo sheets**.

REMEMBER:

Include your research documents with the task you hand to your tutor.

If you cannot do this, take a photocopy of the documents you find, and attach the copy to the task.

You can make your memo formal (using the recipient's full name and title), or informal (using just their name). Be consistent in how you use your name/title.

Remember to sign the memo.

See pages 6 and 7 on how to write memos.

Conducting research and writing memos

Scenario

You work for **Pickton Lift Company** and the company exports its lifts all over the world.

Some staff are going on a sales trip in the last week of next month and it is your job today to find out the weather and temperature in each area being visited.

The staff and destinations are as follows:

Geoffrey Clarke, Sales Manager   Rio de Janeiro, Brazil

Amanda Gregory, Sales Director   Lisbon, Portugal.

Activities

1   Using appropriate reference sources and websites, find and print out the required details for each place being visited.

2   Write a memo to each member of staff attaching the print out/copy of source document relevant to them.

The memo must be headed: Your forthcoming visit to . . . . (put in place name).   Word the memo appropriately.

You are writing the memo and your position is Office Assistant, Travel Department.

> **Remember:**
>
> Your source documents should be included with your task.
>
> Your memo must be laid out correctly and your meaning must be clear.
>
> Correct spelling and punctuation are important too.

TASK 14: HOP, SKIP AND JUMP

Student Information

In this task, you will use information contained in **Appendix 1** and **2**, to complete expense forms for three employees.

You will also make two telephone calls with a partner then complete a **Telephone Message Sheet**.

Ask your tutor for three blank **Hop, Skip and Jump Expense Forms** and a **Telephone Message Sheet**.

REMEMBER:

Be accurate with the information, and calculations, you include on each form.

Complete the forms neatly.

Make notes, before your telephone calls, of what you will say.

See page 17 for guidance on Using the Telephone.

Make notes of what is said during the telephone calls because you will use this information to complete the Telephone Message Sheet.

See page 8 for details of how to take messages.

Use only relevant information on the telephone message form and be sure to be accurate.

Completing expense forms, working with a partner and making telephone calls then completing a telephone message form

Scenario

You work as a Finance Clerk in the Finance Department of a company called **Hop, Skip and Jump**, a company that manufactures shoes based in Your Town.

Today you are to complete several tasks to do with employees' expense forms and hotel bookings. Details of the Approved Hotel rates can be found on *page 50*.

Activities

Working alone

1   **Appendix 1** gives you details of the expenses being claimed for three members of staff as a result of recent business trips.

Complete an expense form for each person.

You can sign as having passed each form for payment.

Note: Staff can only stay in the **Approved Hotel** in each destination. You have a list to refer to when entering details of overnight accommodation.

Working with a partner

2   Your company has now won a major contract in Ipswich and you have to ring **two** hotels – **The River View Hotel** and **The Star Inn** – to find out the rates they charge.

You will need to place **two** telephone calls so you can both take the part of the Hop, Skip and Jump employee, and the hotel employee. One of you will ring The River View Hotel, the other will ring The Star Inn. Prior to making the calls you will need to make notes about what you are going to say in both your roles. Read **Appendix 2** carefully first because it contains some instructions.

Working alone

3   Having got the information you need from the calls, you must complete the telephone message sheet, giving Robert Bouncer, the Finance Officer, the information.

HOP, SKIP AND JUMP

APPROVED HOTELS

HOTEL ACCOMMODATION DETAILS
(Approved hotels and expenses per person per night)

Town/City	Name of Hotel	Cost per person per night £
Chichester	The Star and Garter	76.00
Croydon	The Golden Cross	66.00
Edinburgh	McMurray Hotel	56.80
Exeter	The Westerner	75.35
Huddersfield	Yorkshire Dragoon	45.75
Lincoln	Master and Hounds	62.95
London	Fullingham Lodge Hotel	117.50
Morpeth	The Borderman	53.85
Portsmouth	The Pheasant Inn	73.80
Rotherham	Carlton Hotel	54.75
St Albans	The Coach Inn	84.00
Swindon	The Goddard Arms	94.00
Taunton	Atlantic View	67.45
Wilmslow	The Mancunian	88.50
York	The George and Dragon	75.00

Appendix 1

Expense claim details for last month

Purchasing Division

Matthew Joyce

5th Office to London by train	£107.00	Overnight in London Hotel
19th Office to Lincoln by car	£64.90	Overnight in Lincoln Hotel

Aaron Clifford

19th	Office to Taunton by train	£166.00	Overnight 19, 20 and 21st
22nd	Taunton to London by train	£93.60	Overnight in London
13th	London to office by train	£134.00	No accommodation

Marketing Division

Beverley Grant

4th	Office to Chichester by train	£107.00	Overnight in hotel
5th	Chichester to Office by train	£107.0	
8th	Office to Morpeth by car	£113.55	Overnight in hotel
9th	Morpeth to home by car	£113.55	

Telephone Call Instructions and Guidance

Hop, Skip and Jump Employee	Hotel Employee
You will need to give the name of your company and **possibly** an address and your name and position. Your company anticipates making bookings for around six nights a month – bed and breakfast – one employee staying on each occasion for one or two nights. You wonder if there are special "company" rates? Make it clear staff will pay their own bill upon checkout, i.e. you do not want the bill sent to the company for later payment. You want to know the **latest** check-in time. You wonder how many days notice the hotel will require of a booking, i.e. three days, five days ten days, etc. As your company has been asked to put something in writing, get the address and post code of the hotel.	You will need to identify your hotel and ask for the name of a person to whom you are talking. You will wish to have an estimate of how many bookings per month will be involved. You quote a price of around £57 per night (***there must be at least £10 difference in the cost the two hotels' prices***). Before you can quote an exact rate, you would need to have a request in writing from Hop, Skip and Jump. The latest check-in time is 18:30 unless the hotel receives a telephone call on the day of arrival advising of a later time. The hotel would need to have any booking requested no fewer than three days before arrival.

This is the information you must give or find out. How you do this, the sequence you use and what other information you feel appropriate is up to you.

Do not make each telephone call identical!

TASK 15: VALUE FOR MONEY

Completing a customer refund form

Scenario
You work in the Customer Services Department of **Value For Money Supermarket**. A customer has come to the desk with a complaint concerning being wrongly charged yesterday for tomatoes.

Activities

1   Complete the **Customer Complaint and Refund Form** with the relevant information from the information below.

> Mr Leonel da Silva, yesterday, purchased 2 kilos of Spanish tomatoes priced at 85p a kilo.
>
> His receipt shows he was charged for Cherry tomatoes at a cost of £1.60 per kilo.
>
> The store has decided to refund twice the difference in cost and issue a discount voucher valued at £2. This voucher can be redeemed against any goods in the store in the next 30 days.
>
> Mr da Silva lives at 6 Jackson Street, Your Town BN3 3MX.

Sign the form yourself as the Staff Member, and get a colleague to sign as Mr da Silva.

Use appropriate dates.

TASK 16: PETS SAFE AT HOME

Student Information

In this task, you will design a newspaper advertisement.

Complete an advertising coupon with the wording of your advertisement. Ask your tutor for a copy of the **Advertisement Coupon**.

Write a business letter to a local newspaper asking for the advertisement to be inserted. Ask your tutor for a blank **Pets Safe at Home Letter Heading**.

REMEMBER:

Be brief but accurate with the information you put in the advertising grid.

Write a formal business letter and decide upon a suitable heading for the letter.

You should include an **enc** to indicate something is being enclosed with the letter.

See pages 12 and 13 for advice on writing advertisements.

See pages 14 and 15 for an example of a business letter and *page 16* for useful phrases to include in a business letter.

Designing a newspaper advertisement and writing a business letter

Scenario

You are the Assistant to Bertie Bassett, the Advertisement Manager in the Publicity Department of **Pets Safe at Home** based in Your City.

Your firm offers a "pet sitting" scheme for holiday makers and, as such, has a register of staff upon whom it can call, relating to their particular interest or special knowledge.

The local paper – *The Gazette* – is running a promotional offer with all advertisements at half price for five nights.

Your job today is to design the advertisement and write the letter of confirmation to the Business Advertisement Manager.

Activities

1   Design the advertisement. Remember your aim is to attract custom, so the wording is important. Although the advertisement will be half price, it will be charged per insertion, and your company does not have money to waste.

Complete the Advertisement Coupon (ask your tutor for a copy).

You want prospective customers to realise that you will look, not only after their pet(s), but their home during their period of absence. An additional service is the care and maintenance of greenhouses and indoor/outdoor plants, together with essential food shopping immediately prior to the owner's return.

2   Write to the Business Advertisement Manager at the newspaper and arrange to take up the special £15 single insertion offer, for Monday to Friday nights next week (i.e. five insertions).

> **Note**
>
> This special offer allows for a maximum of 25 words at a charge of £45, before charging another 5p for every word over that amount. (A telephone number counts as one word.) The advertisement size is a quarter-page boxed spread.
>
> **You are not required to calculate the cost, merely to remember WORDS COST MONEY!**

TASK 17: CURRENTLY THE BEST

Student Information	REMEMBER:
In this task, you will conduct some research into consumer law.	Include your research documents with the task you hand to your tutor.
You will then write a personal letter to a store, quoting relevant sections of your researched documents.	*If you cannot do this, take a photocopy of the documents you find, and attach the copy to the task.*
	Write a formal personal letter, including accurate details.
	Use a suitable heading for the letter.
	See pages 18 – 20 for an example how to set out a personal letter.
	See page 16 for some useful phrases.
	Your tone should be appropriate to the task – polite, but firm.

Conducting research and writing a personal letter

Scenario

A number of weeks ago you bought a new Portiflex iron from your local electrical shop.

You discovered the iron was faulty – described as "steam 'n' spray" and did neither!

You returned the iron to the shop but were told by the shop assistant that it was not their problem and you must write to the manufacturer.

Dissatisfied with this you wrote to the Manager of the store and received the reply **shown in Appendix 1**.

Having spoken to a friend, you are vaguely aware that your rights as a consumer are not being taken seriously and decide to do some research then reply to Mr Simonds.

Activities

1   Find relevant information to help you write to the Manager. You should look for information on Consumer Law in the UK, particularly related to the Sale of Goods Act and the Trade Descriptions Act.

Provide copies of your source documents

2   Use the information you obtain to reply to Mr Simonds saying that you will be returning to his store next Saturday and expect to receive a full refund. Make it clear you are not obliged to accept a Credit Note, nor are you willing to accept one on this occasion. You need also to mention it is not your responsibility to contact the manufacturer as your contract is with the store as the seller of the faulty goods.

For each statement you make you must support it with facts.

> **Note**
>
> You will be expected to select an appropriate style and tone and to make your meaning clear, using accurate spelling, punctuation and grammar.

CURRENTLY–THE–BEST
ELECTRICAL STORE

Unit 15, Tilbury Retail Park, Your Town, SK6 3BC
Tel: 01347 4776 2228

(Dated 2 days ago)

Mr/Ms
63 Sunnyville Court
Your Town
SK11 8JS

Dear Mr/Ms

Portiflex Iron Model Spray'n'Steam SuperLuxe

I thank you for your recent letter concerning the purchase of this model of iron for the sum of £35.99. Unfortunately, as my assistant told you when you came into the store, I am unable to help in this matter.

The iron you purchased was sent to us by *Portiflex* and was deemed to be in satisfactory condition and good working order at the time we took delivery of it. As the iron was boxed when you bought it from us it is a matter for the manufacturer to deal with your complaint. I am sure if you return the item to them (address as follows), you will receive a satisfactory outcome to your complaint.

Portiflex (UK) Ltd, 11 Empora Way, Your Town, CH8 15SR

I apologise for this but trust you will realise the matter is out of our hands. You must deal with the manufacturer. Certainly we cannot refund your £35.99, nor can we exchange the iron.

However, as a mark of our goodwill, we are prepared to take the iron back and issue you with a credit note for the sum you paid for the iron. You will have to spend this money in our store within 21 days.

I look forward to hearing from you and wish you success with your complaint to *Portiflex* should you decide upon that course of action.

Yours sincerely

R Simonds

R Simonds
Store Manager

TASK 18: RECYCLING

Student Information

In this task, you will locate, and use, information on **recycling**.

From this research you will prepare:

a) a report;

b) information for a short talk.

You will produce a document to hand to the audience which will contain at least one appropriate image.

REMEMBER:

Include your research documents with the task you hand to your tutor.

If you cannot do this, take a photocopy of the documents you find, and attach the copy to the task.

Make sure that all the information in your talk is accurate.

The image(s) you use will be appropriate and enhance the audience's understanding of the topic.

See pages 10 and 11 for help on using images in communication.

See pages 22 – 26 for help on writing reports.

See pages 27 and 28 for help on giving a talk.

Conducting research and giving a short talk

Scenario

You work in a **The Contented Plaice** fish restaurant, on a part-time basis, and have been concerned, for some time, that it throws away vast quantities of glass and newspaper each day. You mention this to your boss, Margaret Aislaby, and she is impressed by your enthusiasm for, and concern about, the environment. She asks you to put together a report and some information in order to give a short talk to herself and the restaurant's staff.

Activities

1. Conduct some research into the topic of recycling, using the Internet if possible, **and** paper-based sources and write your report.

   You will need to include copies of your source documents and **at least one of these must contain an appropriate image that you will use and explain in your talk**.

   You may wish to consider the following recycling issues:

   - what can be recycled by your restaurant;
   - methods of recycling;
   - the advantages of recycling;
   - facts/figures concerning recycling.

2. Give a short talk about your findings, ensuring you include a document with at least one image, which you will hand to the audience. The talk must last no fewer than four minutes and no longer than six minutes.

Note

Be prepared to answer questions from the audience following your talk.

SAMPLE END ASSESSMENT

20 Multiple-choice questions

The following questions are multiple-choice. There is only one correct answer to each question.

Instructions

1   Choose whether you think the answer is A, B, C or D.

2   Ask your tutor for a copy of the answer grid (or download a copy from **www.lexden-publishing.co.uk/keyskills**).

3   Enter your answer on the marking grid at the end of the test.

4   Hand it to your tutor for marking.

A Communication Key Skills Level 2 External Assessment will consist of 40 questions and you will have **1 hour** to complete them.

How will you select your answers?

If you are sitting your End Assessment in paper format – not doing an online test – you will have to select one lettered answer for each numbered question. The answer sheet will be set in a similar way to the example below:

1   [a] [b] [c] [d]

2   [a] [b] [c] [d]

Make your choice by putting a **horizontal line** through the letter you think corresponds with the correct answer.

Use a pencil so you can alter your answer if you wish and take an eraser to allow you to change your mind about a response. Use an **HB pencil**, which is easier to erase. (If you make two responses for any one question, the question will be electronically marked as **incorrect**.)

Take a **black pen** into the exam room because you will have to sign the answer sheet.

Your tutor has 100 sample End Assessment questions and you will be given these when your tutor considers you are ready to practise the questions.

QUESTIONS

In December 1954, there occurred in London one of the fiercest tornadoes recorded in Britain. A thunderstorm rushed in from the South Coast. The sky turned black an a tornado smashed trees near Hampton Court.	Line 1
Around 5pm the storm reached Chiswick, West London, and there were reports of a huge conical cloud hanging down from the sky, green lightning flashing from its sides and a deafening roar. The tornado hit two factories and tore Gunnersbury station apart.	Line 5
Other damage incurred that day included roofs ripped off houses, chimneys crashing down and walls collapsing.	
A car was reported hurled through the air and windows were rattled violently. The only protection for the terrified people caught outside was to run for cover to try to avoid the barrage of bricks, glass and wood which were thrown through the air.	Line 10
When all was calm and the tornado had passed, news programmes showed a scene of devastation that was described as looking like something from the Blitz.	Line 15
Winds in the tornado's vortex reached an estimated 160kph (100mph) and left a trail of devastation for several miles, finally petering out around Golders Green and Southgate in North London. Remarkably, there were few casualties.	Line 20

Questions 1– 5 refer to the text above.

1   The style of writing in the article is **best** described as:

   A   informative

   B   persuasive

   C   entertaining

   D   technical

2   Which of these words could **best** replace the word **"conical'"**as it is used in **Line 5**?

   A   round

   B   square

   C   tapering

   D   balloon-shaped

3   In the fifth paragraph, which word would best replace the words **"When all was calm and the tornado had passed"** without altering the meaning of the sentence?

   A   Furthermore

   B   Later

   C   Formerly

   D   Remarkably

4   Which of these words could best replace the word **"barrage"** as it is used in **Line 11**?

   A   path

   B   onslaught

   C   whirlwind

   D   number

5   According to the article, which of the following statements is true?

   A   London was well-prepared for the storm.

   B   Very little damage was done but the tornado was unexpected.

   C   There was a lot of damage and some people were killed.

   D   There was a lot of damage but few people were injured.

```
                                                                  33 Swanley Drive
                                                                            Lincoln
                                                                      Lincolnshire
                                                                          LN1 3BY

13 March 2005

Mr J Lines
Customer Services Manager
North Yorkshire Moors Railway
Grosmont
North Yorkshire
NY15 7NP

Dear Sir

Last week, on Tuesday, I traveled on the train leaving Pickering at 10:20, expecting to arrive in     Line 1
Grosmont at 12:25.

When I boarded the train and entered a carriage I found it was strewn with litter and the
windows were dirty on the inside, so much so that it made it difficult to see what was advertised
as "beautiful scenery". When the train arrived at Levisham it was unfortunatley delayed and I did   Line 5
not arrive in Grosmont until 13:05. I, and several other passenger's, tried to establish from
railway officials on the train, the cause of the delay, but our questions were not answered.

Alot of the passengers were annoyed at the delay and I was unhappy about the attitude of the
Station Master at Grosmont when I complained.

I am extremely disappointed that these events and attitudes made what should have been a          Line 10
pleasant day into an unpleasant one and I wish you to look into the matter and let me have your
comments in the next seven days.

Yours sincerely

Colin Crown                                                                                        Line 15
```

Questions 6 – 11 refer to the letter above.

6 The Salutation should have read:

A Dear Mr Crown

B Dear Mr Lines

C Dear Sir or Madam

D Dear Customer Services Manager

7 The underlined word on **Line 1** should be spelt:

A travveled

B travelled

C travled

D travelld

8 The underlined word on **Line 5** should be spelt:

A unnfortunately

B unfortanatly

C unfortunately

D unfortanily

9 The underlined word on **Line 8** should be spelt:

A Allot

B A lott

C All lot

D A lot

10 The underlined word on **Line 6** should be spelt and punctuated:

A passengers

B passengers'

C pasengers

D passengerr's

11 The writer has chosen to start a new paragraph on **Line 10** because he:

A wants to emphasise how much he is annoyed

B is describing what he wants done

C wants to complain about the whole day

D is summarising his feelings and detailing what he wants to occur next

To	Carol Trent	From	Jane Swift	
	Sales Manager		Personnel Manager	

Date   11/10/2004   Line 1

Re   Sunday Trading

In reply to your recent <u>query</u> concerning the hours which are <u>permitted</u> for
Sunday Trading, I can give you the following information. In this country if
retail premises open on Sunday, the maximum trading hours are six.   Line 5
Therefore, we could open at 10am and close at 4pm, or alternatively, 11am
until 5pm. We must not exceed these permitted hours. I trust this information is
helpful. If you wish to discuss this further, please telephone me.

Questions 12 – 15 refer to the text above.

12   This document is an example of:

    A   A telephone message

    B   A letter

    C   A memorandum

    D   A report

13   An alternative for the first word underlined on
Line 3 would be:

    A   question

    B   suggestion

    C   assistance

    D   problem

14   An alternative for the second word underlined on
Line 3 would be:

    A   insisted upon

    B   obligatory

    C   encouraged

    D   allowed

15   The date should have been written as:

    A   11 October

    B   11 October '04

    C   11 October 2004

    D   11/10/2004

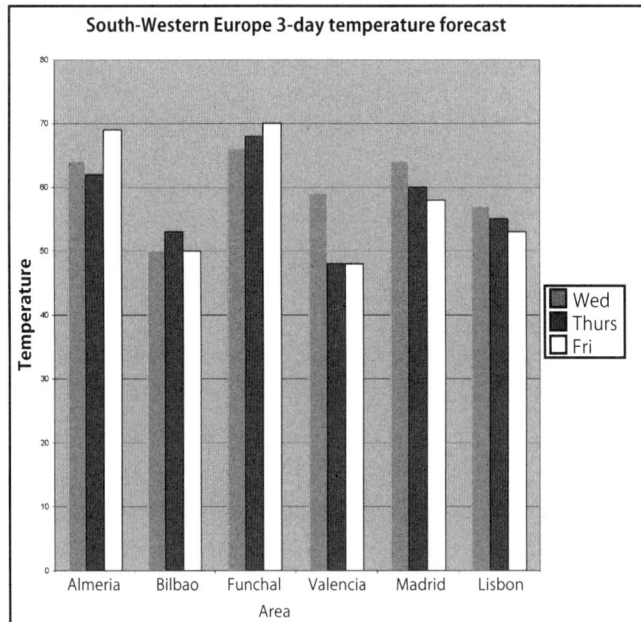

South-Western Europe 3-day temperature forecast

16   On the chart, which **Area** has the highest temperature on Wednesday?

    A   Lisbon   C   Funchal

    B   Almeria   D   Madrid

Outward journey				
From	To	Date	Dept/Arrive	Journey time
Bristol Temple Meads	Exeter Central	Fri 17/12/04	10:14 11:24	1h 10m
Bristol Temple Meads	Exeter Central	Fri 17/12/04	10:40 11:50	1h 10m
Bristol Temple Meads	Exeter Central	Fri 17/12/04	11.16 12:16	1h 00m
Bristol Temple Meads	Exeter Central	Fri 17/12/04	11:20 12:40	1h 10m
Bristol Temple Meads	Exeter Central	Fri 17/12/04	11:44 12:44	1h 00m
Bristol Temple Meads	Exeter Central	Fri 17/12/04	12:16 13:38	1h 22m
Bristol Temple Meads	Exeter Central	Fri 17/12/04	12:44 13:54	1h 10m
Bristol Temple Meads	Exeter Central	Fri 17/12/04	13:44 14:54	1h 10m

Return journey				
From	To	Date	Dept/Arrive	Journey time
Exeter Central	Bristol Temple Meads	Fri 17/12/04	13:50 14:00	1h 00m
Exeter Central	Bristol Temple Meads	Fri 17/12/04	14:00 15:33	1h 33m
Exeter Central	Bristol Temple Meads	Fri 17/12/04	14:46 16:19	1h 33m
Exeter Central	Bristol Temple Meads	Fri 17/12/04	15:00 16:33	1h 33m
Exeter Central	Bristol Temple Meads	Fri 17/12/04	15:10 16:43	1h 33m
Exeter Central	Bristol Temple Meads	Fri 17/12/04	15:19 16:25	1h 06m
Exeter Central	Bristol Temple Meads	Fri 17/12/04	15:45 17:25	1h 40m

Questions 17 and 18 relate to the railway timetable above.

You want to travel to an interview from your home in Bristol to the interview in Exeter on Friday 17 December. Study the timetable and answer the following questions:

17   Your interview starts at 12:30, and you will have a 20-minute walk from Exeter Central station. Which train would you be advised to catch from Bristol Temple Meads?

   A   the 10:14

   B   the 10:40

   C   the 11:16

   D   the 11:20

18   You assume your interview will be over by 13:35 and you want to get back to Bristol as quickly as possible. If the interview **does** end at 13:35, which train would be the **first** you would be likely to be able to catch?

   A   that arriving at 16:19

   B   that arriving at 14:00

   C   that arriving at 15:33

   D   that arriving at 16:43

Questions 19 and 20 relate to apostrophes.

19   Which sentence is correct?

   A   The men's hats' were left on the bus.

   B   The mens' hat's were left on the bus.

   C   The men's hats were left on the bus.

   D   The mens' hats were left on the bus.

20   A   The children's presentation is sheduled for next month. Do you think you'll be able to present the prize to the sucessful child?

   B   The childrens' presentation is scheduled for next month. Do you think you'll be able to present the prize to the successful child?

   C   The children's presentation is scheduled for next month. Do you think you'll be able to present the prize to the successful child?

   D   The childrens presentation is scheduled for next month. Do you think you'll be able to present the prize to the successful child?

Chapter 2: Application of Number

At **Level 2**, a learner should be able to **add**, **subtract**, **multiply** and **divide** with little or virtually no problems.

However, at **Level 2**, problems may be substantial and may require calculations involving two steps or more. In other words, breaking a problem down in to smaller manageable parts, to form a solution. For example, using a combination of addition followed by a multiplication, and finally a division to find a solution. Combinations may vary.

The following Reference Sheets show the numeracy functions a learner should be able to use effectively. It must be stressed that numeracy problems can be a combination of functions and problems, and as a learner you should be able to select appropriate methods to obtain the correct results. You are expected to be able to demonstrate your ability to **collect** information, **process** that information and **interpret** the information.

UNDERSTANDING NUMBERS

Whole numbers

Read, write, order, positive/negative, estimate, compare, use.

Word	Number
one	1
ten	10
one hundred	100
one thousand	1,000
ten thousand	10,000
one hundred thousand	100,000
one million	1,000,000

A lottery winner won **£2,456,125.19** – this could be written or read as **two million, four hundred and fifty-six thousand, one hundred and twenty-five pounds and nineteen pence**. Remember to start with the largest number through to the smallest number.

There is also the **ordering** of numbers to consider, for example:

Numbers	
1st	first
2nd	second
25th	twenty-fifth
50th	fiftieth
100th	one hundredth
500th	five hundredth

Other examples can be dates such as **23rd May 2001**.

Recognising positive/negative numbers

Positive numbers and negative numbers can be such things as currency, temperature, time/date or units of measure, etc.

The simplest way to understand positive and negative numbers is to use a **number line**. Negative numbers always have a minus sign (-) before them, i.e. **-5° C** or **-£15**. Positive numbers don't usually display the + sign before them. Zero is always in the middle. For example:

Temperature

Bank balance

The above diagram shows both temperature and a bank balance for comparison.

A numberline can also be vertical. The choice is yours.

Rules for adding and subtracting

Adding a negative number is the same as subtracting. For example:

7 + (-4) is the same as **7 - 4 = 3**

Subtracting a negative number is the same as adding. Two negatives make a positive. For example:

(-5) - (-3) is the same as **(-5) + 3 = -2**

Rules for multiplication and division

positive $\times$ positive = positive   (e.g. 7 x 7 = 49)

positive $\times$ negative = negative   (e.g. 7 x -7 = -49)

pegative $\times$ positive = negative   (e.g. -5 x 5 = -25)

negative $\times$ negative = positive   (e.g. -9 x -4 = 36)

When using division follow the same rules as multiplication. For example:

$-36 \div 6 = -6$ (because $36 \div 6 = 6$ and the answer must be negative)

$25 \div -5 = -5$

$-48 \div -6 = 8$ (the same as multiplication)

There follows some examples of positive/negative numbers that you may do in the Level 2 End Assessment.

Q   A fridge has a temperature of **2°C** and a freezer has a temperature of **-9°C**. What is the difference in temperature?

✓   Think of, or draw, a numberline and work out the difference. Equals **11°C**.

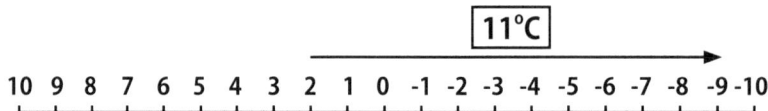

$$\boxed{11°C}$$

| 10 | 9 | 8 | 7 | 6 | 5 | 4 | 3 | 2 | 1 | 0 | -1 | -2 | -3 | -4 | -5 | -6 | -7 | -8 | -9 | -10 |

Q   Inside an aeroplane the temperature is **22°C**, but outside it is **-40°C**. What is the **difference**, or **spread**, of temperature?

✓   There is **22°C** on the **positive** side and **40°C** on the **negative** side. Add the two temperatures (**22 + 40**) to get a difference of **62°C**.

Q   A business has an income during a month of £100,000. However, costs were £50,000 and capital outlay was £200,000. How much money does the business have this month?

✓   First, separate income (which can be thought of as profit) from costs and capital outlay (which can be thought of as expenditure). Then treat it as a simple minus sum.

Profit		Expenditure
	Difference	£50,000
£100,000		£200,000+
= £100,000	-Take away -	= £250,000

Finally, £100,000 **minus** £250,000 will leave **-£150,000**.

Estimation

Estimation is using your best judgement or approximation.

You can use **estimation** to say how much fluid is in this beaker:

You can estimate the beaker as being **half full**. It can be also be said that it is **half empty**.

Look at this tape measure:

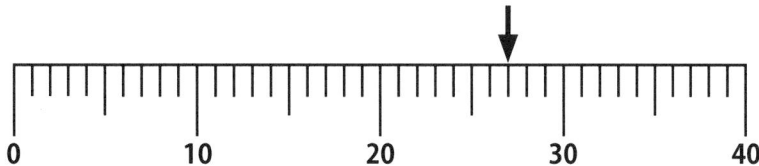

After looking at this tape measure it can be estimated that the length indicated by the arrow is **approximately** 25 units. It can also be said it is **approximately** 30 units.

Q Three items from a shop cost £31.80, £10.05 and £4.99. Estimate the total cost?

✓ Simplify the amounts to the nearest **whole pound**. Then calculate an answer:

£32 + £10 + £5 = £47

Q A worker charges £30.20 per hour and takes 21 hours to complete it. Estimate the final bill?

✓ Make **£30.20 = £30** and **21 hours = 20**. So **£30 multiplied by 20 = £600 final bill**.

Q A boy has maths homework to do. He has 4 rows of problems.
There are 9 problems in each row. If he can do 10 problems every 15 minutes, estimate how long will it take for him to finish his homework?

✓ You can estimate **9 as 10**, and 4 rows by 10 problems = **40 problems**. So 10 problems every 15 minutes would mean **15 x 4 = 60mins**. Answer = **1hr** for his homework.

Rounding off

Rounding off is a way of simplifying numbers. For example, if you won **£3,673,017.12** it would be easier to write **£3,500,000** or **three and a half million pounds**. Rounding off numbers also makes it easier if you want to estimate. Rounding numbers to the nearest 10, for example, means finding which 10 they are nearest to. Rounding a number to the nearest hundred, to the nearest thousand or even bigger can be done in the same way.

Rules for rounding off

If you were asked to round off **145** to the nearest **10** it would either become **140** or **150**. Look at the following rules:

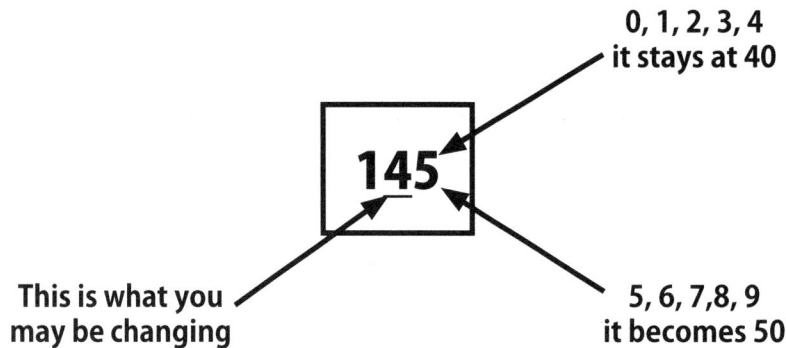

0, 1, 2, 3, 4
it stays at 40

145

This is what you may be changing

5, 6, 7,8, 9
it becomes 50

(Q) A pencil is 27cm long. How long is it to the nearest **10cm**?

(✓) You are looking to possibly change the **10cm** unit. Look at the **7 unit** beside the **20 unit** as this will affect changing this up one unit or staying the same. As it is **7** we must round up to **30cm**.

(Q) A sprinter runs 1,400 metres around a track. What is this to the nearest one **thousand** metres?

(✓) You are looking to possibly change the **1,000m** unit. Look at the **4** beside the one thousand metre unit. This means that the one thousand metre unit will not change, so the answer is **1,000m**.

Accuracy

Accuracy is used to define an acceptable level. For example, if you are completing a long multiplication sum you may be required to give your answer to **two decimal places accuracy**. Your answer may then be **19.23** to two decimal places.

In some respects it is **similar** to rounding off as it also uses given levels of accuracy.

Other examples of where accuracy may be used are in **measuring** and **time**.

You may **measure** the length of a room and use an accuracy of one centimetre rather than being precise to the exact millimetre. But if you were measuring to fit a new door frame an accuracy of one millimetre would then be necessary.

If somebody asked you for the time a usual reply might be, "nine-thirty" or "twenty to eleven". Giving the time to the nearest second would generally not be used. However, a more precise and accurate time would be acceptable in a 100 metres race where the difference of one hundredth of a second is crucial.

Ratio

A ratio is a way of proportioning numbers or quantities into **parts**. To put it another way, ratios are about **sharing items out**. It can be as simple as mixing paint colours such as, red and black to get another colour.

Ratio is also another way of comparing the relationships between two or more quantities such as in scale diagrams.

At **Level 2** you should be able to work out ratio to at least three parts and recognise that ratio can be displayed as: **1 to 20**, **1 : 20**, **1/20**. You must also recognise ratio can be used for scaling objects up or down.

Mixing paint is an ideal example of ratio. Consider the following:

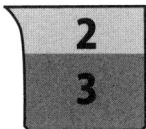

In this mixing pot of paint there are **2 parts** of one colour and there are **3 parts** of another colour.

Importantly, you must remember that there is a **total of 5 parts**.

If we were to make up **5 Litres** of paint we would need **2 Litres** of one colour and **3 Litres** of another colour. However, it is usually not that easy.

If we require **20 Litres** of paint with the ratio **2 : 3** yellow to blue, how much of each colour will we need?

First, we add the **2 + 3 = 5** as this shows we have **5 parts** in **total**. Next we **divide 5 in to 20 Litres** as this will show the size of **one part** of the **5**. So **20 ÷ 5 = 4**.

Next we use this answer of **4** and multiply it to each colour:

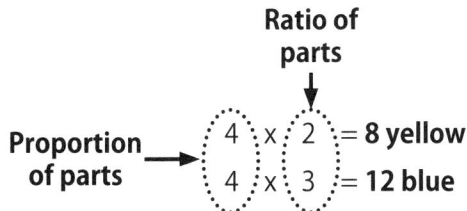

Ratio of parts

Proportion of parts → 4 x 2 = **8 yellow**
4 x 3 = **12 blue**

Let's use a check calculation to **prove** this works by adding **8 + 12 = 20**. So we have 8 Litres of yellow to 12 Litres of blue (2 : 3).

Another example would be to divide an amount of £400 in to a ratio of **5 : 3**

First add 5 + 3 = 8 (parts) then

£400 ÷ 8 = £50 (for 1 part of the total)

So, £50 x 5 = £250 (5 parts of total)

£50 x 3 = £150 (3 parts of total)

£400 in a ratio of **5 : 3 = £250 to £150** (check calculation **250 + 150 = 400**).

Q There are 2,000 employees in a ratio of **3 : 7 males to females**. How many males are there?

✓ Add 3 + 7 = 10 Next divide 2,000 by 10 equals 200 (**1 part**) Then 3 (males) x 200 = 600. There are **600 males**.

Q An amount of money is divided in a ratio of **3 to 2**. The smaller amount is £40. What is the total amount of money originally?

✓ The £40 = two parts of the ratio, so **£40 ÷ 2 = £20** and this is one part of the total. Now there was the other part of the ratio which was **3**. We **multiply** that by **£20** to get **£60**. Finally, the total amount was **£100 as £60 + £40** was the original ratio of **3 to 2**.

Scale

Scales use ratio in a similar way. If you photocopy a picture by **1 : 1** it will come out the **same size**. But if you were to photocopy it at **1 : 2** it will come out **half the size (1/2)**. However, you may want to enlarge the picture by **2 x**. This will be a scale of **2 : 1** making the picture twice as big. For example:

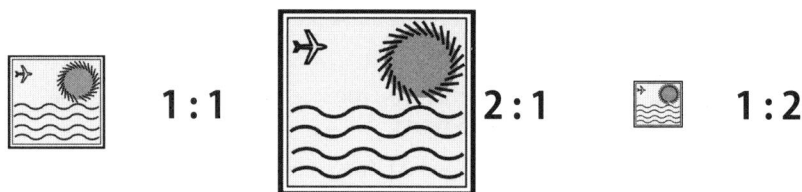

At **Level 2** you should be able to use and understand scales in things like diagrams or maps.

Q If you made a model boat to the scale of **1 : 22** and the original was **40 metres** long, how big would the model be?

✓ First, remember that the model will be **22 times smaller**. So divide **40 by 22**, **40 ÷ 22 = 1.82m**, to an accuracy of 2 decimal places.

Q A carpenter uses a scale of **1cm to 2m** for a kitchen plan. How big would a kitchen unit be if shown on the drawing as **1.5cm by 0.5cm**?

✓ Since **1cm = 2m**, multiply **1.5 by 2 (1.5 x 2)** this equals **3**. Next multiply **0.5 x 2** equals **1**. The life size dimensions of the kitchen unit will be **3m by 1m**.

Q Look at the map. What is the distance from Uptown to Weston using a scale of **1cm to 0.5 miles**?

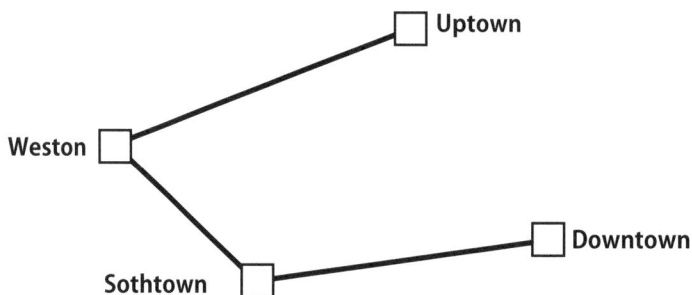

✓ You would be provided with a ruler to measure the distance between Uptown and Weston. **For every centimetre the distance is equal to 0.5 miles in real life**.

Proportion

Proportion is when two or more quantities increase at the **same rate** as each other. For example, the further the distance that you travel in a car, the more fuel you are going to use. From this you can see a **direct relationship**.

Other examples include – the more hours you work in employment the more money you will be paid in wages. Note also, the more you earn the more in tax you are likely to pay.

The reverse is also true and is known as **inverse proportion**. For example, it takes two men six days to paint a house. If there were four men painting then it should take three days to paint the house.

Q An engineering company employs **10 people to produce 500** car parts over **10** days. How long would it take 15 people to produce the parts?

✓ The people employed to produce the 500 parts has been increased by **50%** (**1/2**), so the time to produce the parts should be reduced by the same amount. So **half** (50%) of 10 days equals 5 days to produce the parts.

Fractions

Understand, order, increase/decrease, compare.

Fractions are **parts of a whole**. A **half** (1/2) of **one** (1) is a **half** (1/2), and **two halves make one** (1/2 + 1/2 = 1).

At **Level 2** you should be able to use and understand simple fractions, and also see their relationship to decimals and percentages.

Fractions are made up of **two parts**, a numerator and a denominator. Below is a diagram to show the relationship between them:

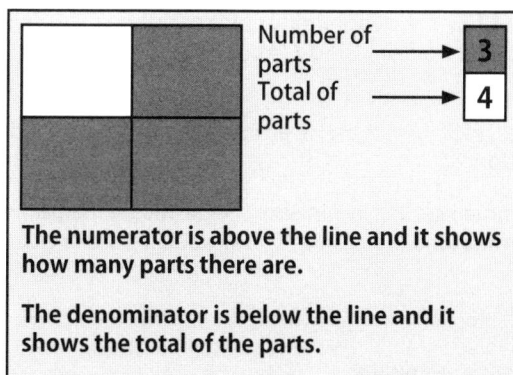

Number of parts → **3**
Total of parts → **4**

The numerator is above the line and it shows how many parts there are.

The denominator is below the line and it shows the total of the parts.

It is important to know the size and relationship fractions have to each other. The following chart shows some fractions as an **image**; the way a fraction is **written** and **read**. Included within the **Read** section is an explanation.

Model	Fraction	Read
	$\frac{1}{2}$	One half, one out of two, one divided by two.
	$\frac{1}{4}$	One fourth, one out of four, one divided by four.
	$\frac{2}{3}$	Two thirds, two out of three, two divided by three.
	$\frac{3}{5}$	Three fifths, three out of five, three divided by five.

It may be easier to think of fractions by using a **numberline**. Here is an example using some popular fractions:

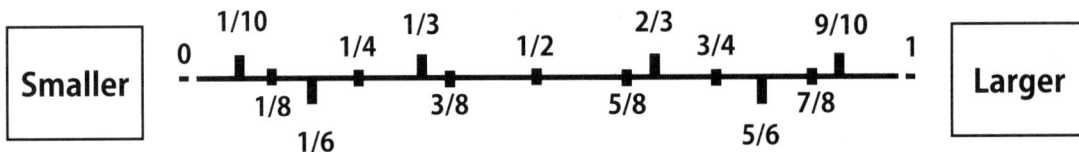

Fractions sometimes need to be **simplified**. For instance **2/10** can be simplified because 2 goes into 10 five times and into itself once, giving an answer of **1/5**. Here is another example:

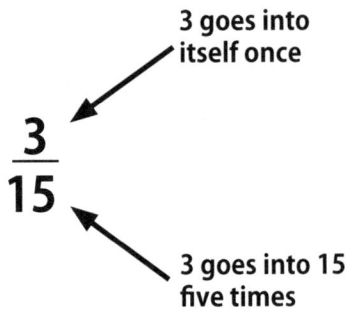

3 goes into itself once

$$\frac{3}{15}$$

3 goes into 15 five times

Answer = **1/5**

Q What is **3/4 of £2**?

✓ First divide the quantity **£2** by the **denominator** of the fraction (2 ÷ **4** = 0.5) then multiply the answer by the **numerator 3**. This is **3** x 0.5 = 1.5. So **3/4 of £2 equals £1.50.**

Q A man is paid, after taxes, £180 in wages. He estimates that **1/3** was deducted in tax. What was his full wage before tax deductions?

✓ **2/3 equals £180** (remember he's lost 1/3 in tax). Divide 180 by the **numerator 2**, equals **90**. 90 is then multiplied by the **denominator 3**. 3 x 90 = 270. The wage before deductions was **£270.**

Decimals

Understand, order, increase/decrease, compare.

A decimal is a number that uses place values and a decimal point to show amounts that are **more** than or **less** than one, such as zero point two five (**0.25**) or **£5.95**.

At **Level 2** you should be able to work with decimals using basic arithmetic and compare/convert to fractions and percentages.

Importantly, you must be able to order decimals from **largest** to **smallest**. For example:

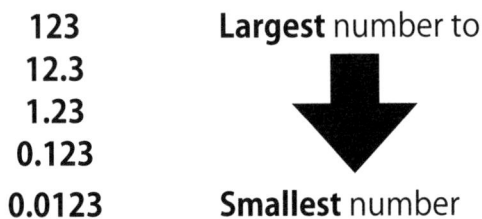

123	**Largest** number to
12.3	
1.23	⬇
0.123	
0.0123	**Smallest** number

You may want to use a **numberline** to help order decimals, such as this one:

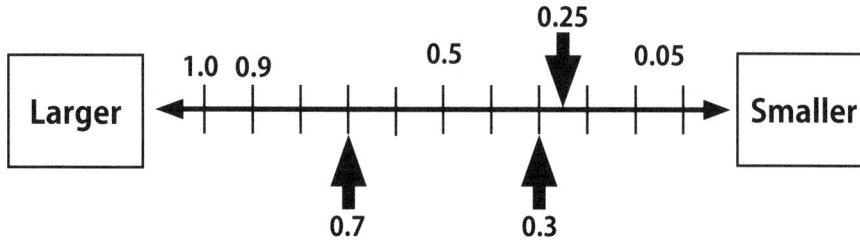

Q Place the following decimals in to order **largest to smallest**:

2.9
1.9
1.09
2.09

✓ First look to the **left** of the decimal place and decide which is the **largest number**. Then from the **right** of the **decimal place** in the **first column**, decide which is the next largest. Use the same technique again moving to the right from the previous column. Placed in order of size the decimals will be:

2.9, 2.09, 1.9, 1.09

Q How do you multiply **3.75 x 9.2**?

✓ This should be treated the same as a **long multiplication** sum first. Look at the following method:

```
    3.75  ←
  x 9.2   ←      Three numbers
   750           to the right of the
 33750 +         decimal place
 ───────
 34.500
```

**Decimal place reinserted
three places from the right** →

After the multiplication sum is **complete** the decimal place is **reinserted**. This is done by **adding** up the number of figures used to the right of the decimal place. **In this case there are three**. The decimal point is placed back in **three places from the right** in the final answer as shown.

Percentages

Understand, order, increase/decrease, compare.

At **Level 2** you will be expected to calculate and use percentage values, and also convert to fractions and decimals.

Percent means **out of a 100**. It is a set of number values usually up to a maximum value of **100%**. A good way to show percentage values is to use a **numberline.** Here is an example:

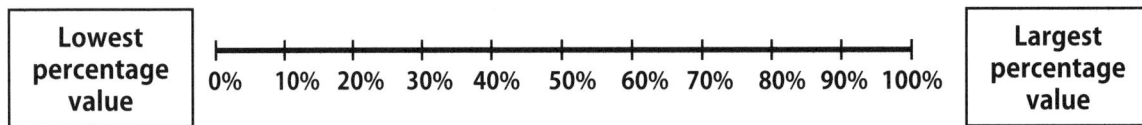

Lowest percentage value			Largest percentage value

0% 10% 20% 30% 40% 50% 60% 70% 80% 90% 100%

100% means the **full value** of a figure. **50%** means **half** of the value, for example a music CD normally selling for £12 would be £6 if it was on sale at **50% off**.

But how could you work out 50% off on a CD costing £9.72?

The usual mathematical method would be:

$$\frac{£9.72}{100} \times 50 = £4.86$$

Another method is to multiply **50 x £9.72 = 486** Then divide the answer by **100**, which equals **£4.86**.

And another method is to find **10%** of £9.72 by moving the numbers one place to the right of the decimal point = **0.972**. Next, multiply this answer by **5 = £4.86**.

These are different methods that give the same answer. The method used is your choice.

Q At a sporting event there are a **total of 1,200** people. **300** of these were female spectators. **What is the number of female spectators as a percentage of the 1,200?**

✔ If you were using a calculator, then **300 ÷ 1,200 x 100 = 25%**
Another method would be to realise that 1,200 is the full number and equals 100%. If you halve 1,200 (**which is the 100% figure**) it would equal 600 (**which is 50%**), this can be halved again, 600 (50%) would equal 300 (which is 25%). **The females would equal 25% of the total**.

Q The percentage of profit for selling a computer is worked out by using the following formula:
(£S – £P) ÷ £P x 100

If **£S** equals the selling price of **£600** for the computer, and **£P** is the purchase price of **£400**, what is the percentage profit?

✔ Because it is a formula, the figures need to be inserted and then calculated.
(£S – £P) ÷ £P x 100 with the figures in: **(600 – 400) ÷ 400 x 100**, this is **200 ÷ 400 x 100 = 50**. So the profit equals **50%**.

Q A car normally retailed at £10,000 in 2004, but was increased to £10,500 in 2005 due to an upgrade in specifications. What was the percentage increase between the two figures?

✔ Calculate the difference between the two figures, **£10,500 - £10,000 = £500 difference**. This figure is used with the original figure in a percentage calculation now, but before this is done the figures can be simplified. **500** can become **5** and **10,000** can become **100**.

$$5 ÷ 100 \times 100 = 5$$

So there was a **5%** increase between 2004 and 2005 car prices.

Fractions, decimals and percentages

As a **Level 2** student you must be able to understand the **relationship** between fractions, decimals and percentages. Again a **numberline** is useful to note this relationship. Here is an example:

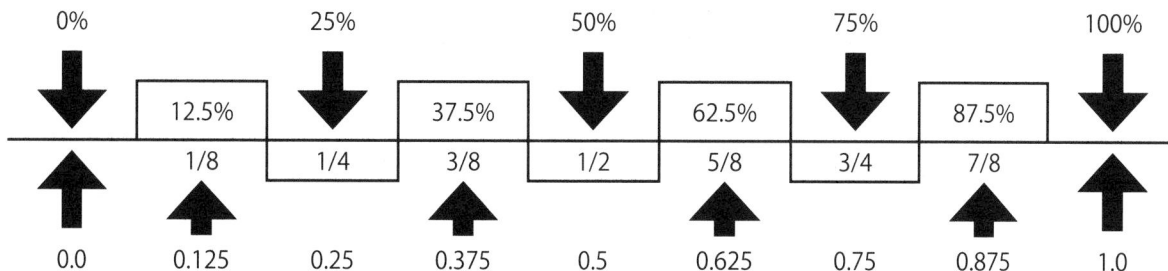

0%		25%		50%		75%		100%
	12.5%		37.5%		62.5%		87.5%	
	1/8	1/4	3/8	1/2	5/8	3/4	7/8	
0.0	0.125	0.25	0.375	0.5	0.625	0.75	0.875	1.0

Q **Forty** athletes are trying to get on the national team. **25%** initially fail the first trial. At the second trial **1/3** fail. Then at the final trials, only **0.5** pass. **How many athletes make it on to the national team?**

✓ From the original **forty** 25% fail, this leaves us with **30** because 25% of 40 = 10 (**4 x 10 = 40**). At the second trials **1/3** of thirty now fail, this leaves us with **twenty** (**1/3 of 30 = 10, or 3 x 10 = 30**). Now at the final trials only **0.5** pass from **20**, this leaves us with ten (**0.5 x 20 = 10**). So **there are only 10 who make the national team**.

Q What is **3/8** as a **decimal**?

✓ To convert **3/8** in to a decimal **divide 3** (numerator) **by 8** (denominator) **3 ÷ 8 = 0.375** (**here is the long division working out**) you may work it out differently though.

$$
\begin{array}{r}
0.375 \\
8\overline{)3.000} \\
\underline{24} \\
60 \\
\underline{56} \\
40 \\
\end{array}
$$

Q What is **0.25** as a **fraction**?

✓ Decimals are easily converted into fractions. **0.25 is a quarter of 1** or in other words, there are four 0.25s in 1. So **0.25 = 1/4**. Using a numberline of decimals and fractions is useful here.

Q What is **20%** as a **fraction?**

✓ Again possibly using a numberline may help to see the relationship of percentages and fractions. Alternatively, put the **20%** over the **100%** as this is the fraction **20/100**. Now the fraction needs to be simplified. **20 goes into itself once and into a 100 five times giving an answer of 1/5**.

Using a calculator

At Level 2 you should be able to use a standard or simple scientific calculator competently using the basic arithmetic functions, squaring values, percentages, decimals, negative numbers, and fractions and using π in a range of calculations.

You are **not** allowed to use a calculator during the End Assessment. However, you are encouraged to check and use a calculator during Part A Tasks and when completing your portfolio. **However, you should not rely on a calculator for every task**.

UNITS OF MEASURE

Understand units of measure, order, increase/decrease, compare.

The **metric system** is a system of measurement that describes the measurement of an object, its weight, or its volume. It uses the decimal system in powers of ten. The units are measured with rulers, tapes, scales, etc.

There is also the **imperial measuring system**. This is still occasionally used today in certain areas of employment and examples are given on *page 75*.

At **Level 2** you will be expected to be able to use the metric system and its various units and also the imperial system if required. You will also need to be able to convert or compare between the two systems as necessary. A formula may be provided for converting units of measure.

The metric system

Length is measured using the units of:

		millimetre (mm)
10mm =	1cm	1 centimetre (cm)
100cm =	1m	1 metre (m)
1,000m =	1km	1 kilometre (km)

The following are examples of measurements in length:

8mm =	0.8cm
12mm =	1.2cm
435mm =	43.5cm
37cm =	0.37m
87.5cm =	0.875m
490m =	0.49km
3,590m =	3.59km

Weight is measured using the metric units of:

	grams (g)
1,000g =	1 kilogram (kg)
1,000kg =	1 tonne

The following are examples of weight:

250g =	0.25kg
2,755g =	2.755kg
1.25kg =	1,250g
4.8kg =	4,800g

Capacity is measured using the units of:

	millilitre (ml)
10ml =	1 centilitre (cl)
100cl =	1 litre (L)
1,000ml =	1 litre (L)

The following are examples of capacity:

$$30ml = 3cl$$
$$200ml = 20cl$$
$$750ml = 0.75L$$
$$5,500ml = 5.5L$$

The imperial system

The imperial system of measuring uses its own units of measurement.

At **Level 2** you should understand the imperial system and convert to the metric system with given formula if necessary.

Length in the imperial system:

$$inches\ (")$$
$$12\ inches = 1\ foot\ (ft)$$
$$3\ feet = 1\ yard$$
$$1,760\ yards = 1\ mile$$

Examples of length:

4 inches

1ft 3 inches

1½ miles

Weight using the imperial system:

$$ounces\ (oz)$$
$$16oz = 1\ pound\ (lb)$$
$$14lb = 1\ stone\ (st)$$
$$8st = 1cwt$$

Examples of weight:

6lb

2½lb

4¼st

Capacity in the imperial system:

$$20fl\ oz = 1\ pint$$
$$8\ pints = 1\ gallon$$

Examples of capacity:

4 pints

9 gallons

Measurements in the **metric** and **imperial** system can be taken using various instruments such as scales, tapes, jugs, digital appliances, etc. At **Level 2** you will be expected to read these instruments accurately, and to given levels of accuracy, in whatever system it is measuring.

The following is an example of a scale using both **metric** and **imperial** systems:

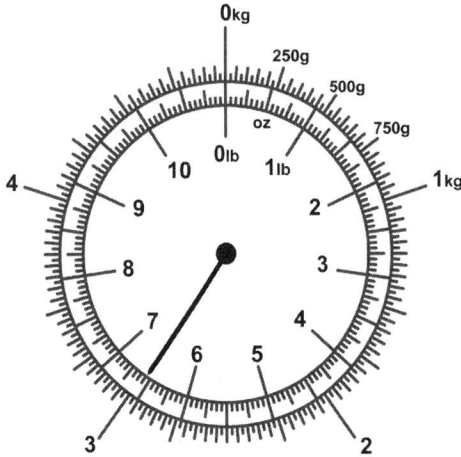

This scale shows the various weight scales in comparison to each other.

The weight is slightly over 6 1/2lb and just under 3kg.

Time

Time is measured in different units and can be measured either **digitally** or by an **analogue** clock using the **12-hour** or **24-hour** format.

At **Level 2** you should be able to use the **12/24-hour** format and work using seconds, minutes, hours and days.

Example of time units used:

60 seconds =	1 minute (min)
60 minutes =	1 hour (hr)
24 hours =	1 day
7 days =	1 week (wk)
12 months =	1 year (yr)
1 year =	365 days

The following are examples of time used daily:

1 min 30 secs

4 hrs 45 mins

48 hrs

2 days 9 hrs

Here is an example of a digital clock showing 19:05 hours. It shows the hours and minutes. The **am** and **pm** may also be shown:

This can also be displayed as **7:05pm** in the 12-hour format.

Here is an example of an analogue clock showing **10:10**. It could also be showing **22:10** in the 24-hour format:

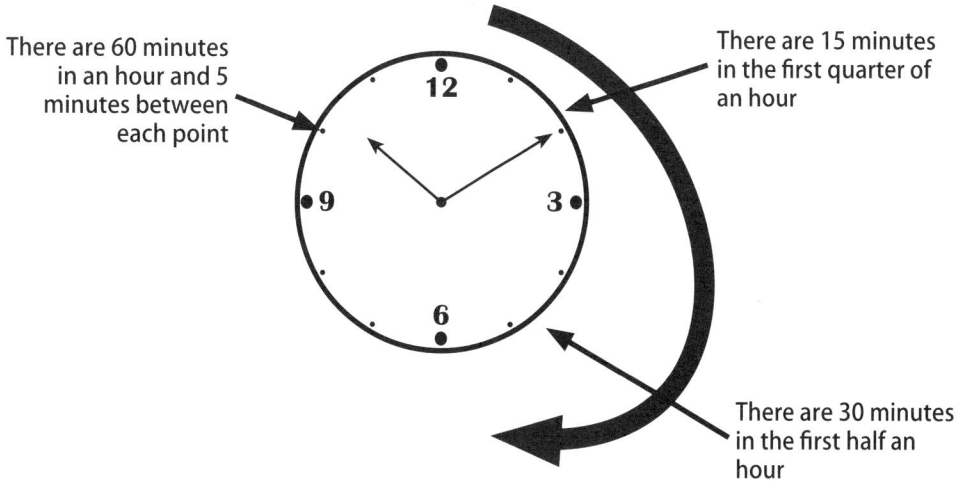

There are 60 minutes in an hour and 5 minutes between each point

There are 15 minutes in the first quarter of an hour

There are 30 minutes in the first half an hour

Q An office worker takes approximately **5 minutes** to type **140 words**. How long will it take for the worker to type **4,200 words**?

✓ 4,200 words **divided** by 140 words, this equals 30. Then **5** minutes multiplied by **30** equals **150** minutes. **This can also be written as 2 hrs 30 mins.**

Q The ferry crossing from England to Ireland leaves at **22:30 hrs** and takes **2hrs 45 mins**. What time will it arrive?

✓ First **add 2 hrs** on to **22:30 hrs**, equals **00:30 hrs**. Then add the remaining **45 mins** to this, equals **01:15 hrs**. It may help to draw a clock face in more complex problems to aid finding the answer.

Temperature

Temperature is generally measured in **degrees Celsius** and shown by the symbol **°C**. There is also another measure of temperature and that is in **degrees Fahrenheit** and this is shown by the symbol **°F**. Degrees Celsius tend to be the most widely used.

At **Level 2** you must be able to convert between the two units of measure using given formula.

Temperature can be measured by a thermometer or digitally. It is important to realise that a temperature may go **below 0** and this makes it a negative number. For example, it can be said that the temperature outside is **5°C below 0**. This is the same as **-5°C**.

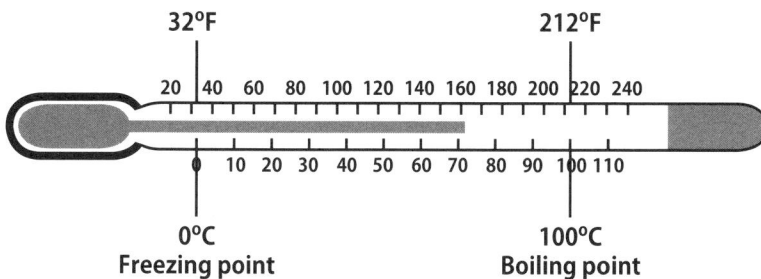

32°F 212°F

20 40 60 80 100 120 140 160 180 200 220 240

10 20 30 40 50 60 70 80 90 100 110

0°C
Freezing point

100°C
Boiling point

This is an example of a thermometer showing both **Celsius** and **Fahrenheit**.

Notice that the scales are different. Boiling point is **100°C** or **212°F**.

Using a digital scale temperature will look like this:

$$37.5°C$$

Q The temperature has **dropped** five degrees during the night from **3°C** during the day. What is the temperature during the night?

✓ Using a **numberline** here is useful. Work backwards from

$$\xrightarrow{\quad 5 \quad}$$
3, 2, 1, 0, -1, -2

and we get an answer of **-2°C**.

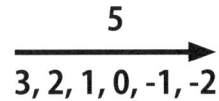

Q Using the following formula calculate **25°C** in to **°F**:

$$F = (C \times 9 \div 5) + 32$$
Where **F** = degrees Fahrenheit and **C** = Celsius

✓ First insert the numbers that we have in to the formula **F = (C x 9 ÷ 5) + 32**.
This becomes :

?°F = (25°C x 9 ÷ 5) + 32. Complete the bracket calculation first **?°F = (45) + 32**. Then **45 + 32 = 77**.

So the answer to this temperature question is **77°F**.

SHAPE AND SPACE

Find area, perimeter and volume.

At **Level 2** you should be able to work out area, perimeter and volume from given dimensions or formula.

Area

Area is calculated by length multiplied by width (**L x W**) and shown by the symbol ^{2} (**squared**). **This is actually a simple formula where letters are replaced by numbers**.

At **Level 2** area calculations are generally based on **L shaped** dimensions. The process may involve working out missing dimensions to find the area. For example, look at the storeroom illustrated below:

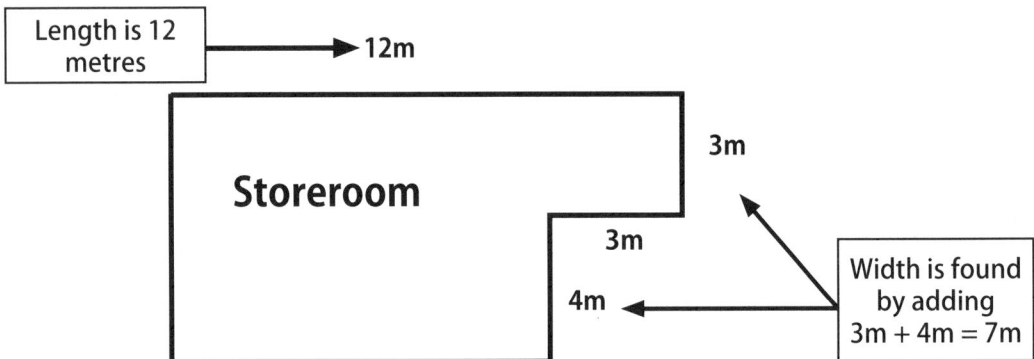

The **general area** can be found by multiplying **12m x 7m = 84m²**. But this is not the correct area. There is the **cut out** of **4m x 3m = 12m²** to be taken away from the general area total. So **84m – 12m = 72m²** for the area of the **L shaped** storeroom. *See Diagrams, page 86 and 87 for further examples.*

Sometimes you may have to find the area of circles. The formula used is always given including the value to use for π **and this is nearly always 3**. For example:

Area of a circle = πr^2

$\pi = 3$

r = radius

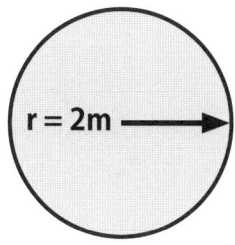

Area = πr^2

First, put the numbers into the formula

Area = 3 x 2m² (remembering to multiply)

The circle area = 12m²

Q What is the area of this house shown by the plan view?

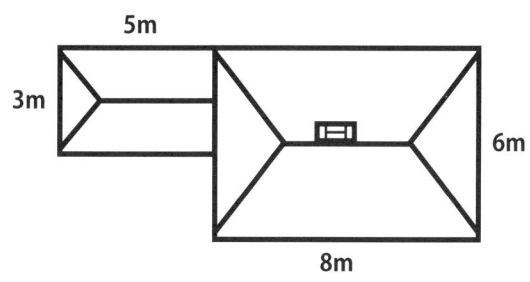

✓ There are **two** simple **oblong** areas to work out then add together to get a total. So the first one is **3m x 5m = 15m²** then, **8m x 6m = 48m²**. Then add the two together to get the **total area 15m² + 48m² = 63m² total area**.

Perimeter

Perimeter is the distance **around an object** and is found by adding all the dimensions together to form a **total**. If we were to find the perimeter in this example we would have to work out the **missing** dimensions first.

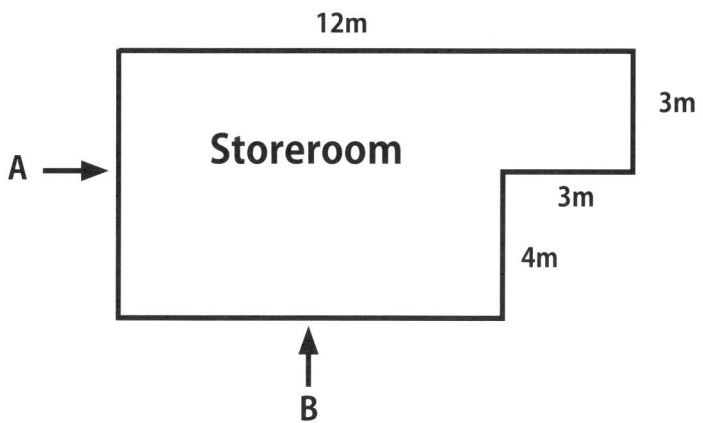

The value of **A** is found by adding the dimensions 3m and **4m** together (**7m**) as they are **parallel** with **A**. Then the value of **B** is found by taking **3m** away from **12m** (**9m**).

So the **perimeter is 12m + 3m + 3m + 4m + 9m + 7m = 38m**.

Volume

The volume of objects is generally found by multiplying the **length x width x depth** and is shown by the symbol ^{3}. Sometimes a formula may be given. For example:

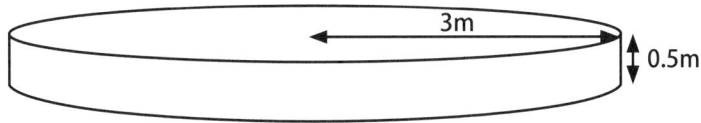

The volume of this object is found by:

$$\pi r^2 d$$

$$\pi = 3$$

The value for π has been given as **3** and the other dimensions are shown. So, we insert the numbers into the formula $\pi r^2 d$, which becomes **3 x 3m^2 x 0.5m**. This equals **13.5m^3**.

Q   What is the volume of the following object?

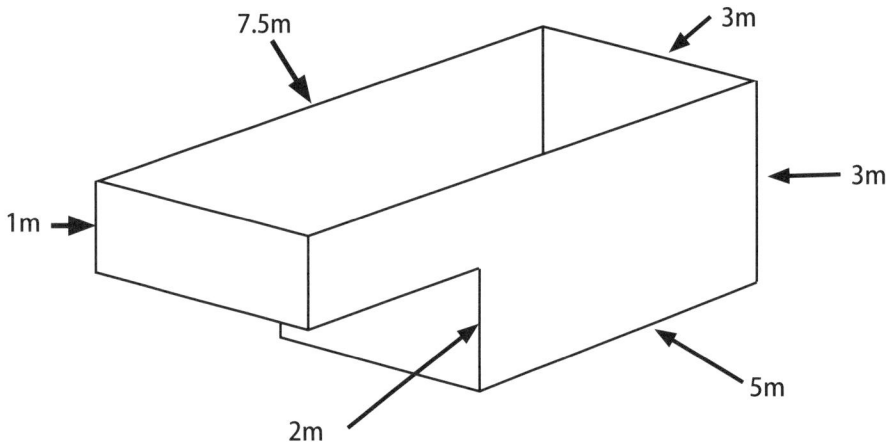

There are a number of ways to find the volume of this object. One way is to find the area of the side of the object (L shape) then multiply that by the depth (3m). Another way is to split it into two box shapes (small + large), then add the two volumes together.

Using the first method, part of the area is **5m x 3m = 15m^2**, then add this to **1m x 2.5m = 2.5m^2** which equals **17.5m^2**. Now this is multiplied by **3m (3m x 17.5m) = 52.5m^3**

USING FORMULAE

At **Level 2** you should have an understanding of given formulae.

A formula may be something as simple as finding area, **L x W** where **L** is length multiplied by **W** width. It can also be in the volume of a cube, **L x W x D**. It is also used in finding the area of a circle by πr^2 or in the volume of a cylinder, $\pi r^2 d$.

Whatever formula is given for you to use, it is simply a case of replacing **letters** or **symbols** with **numerical values**.

Q A printer calculates a customer's bill by using this formula **C = F + (0.05n)**.

C is £s charged to customers

F is the set up cost of £45

n is the number of prints

How much will it cost to print **500** copies?

> **Note**
>
> Expressions written like **CF** or, in this instance **0.05n**, mean you are to multiply them together, i.e. C x F or 0.05 x n.

Insert the numbers into the given formula and calculate:

C = 45 + (0.05 x 500)

This equals **70**. So the printer charges the customer **£70**.

Q The volume of a can is found by using $\pi r^2 d$. Look at the following and calculate the volume.

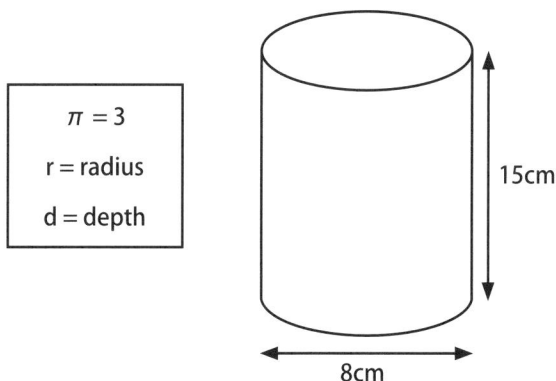

$$\pi = 3$$
$$r = \text{radius}$$
$$d = \text{depth}$$

15cm

8cm

Insert into the formula the figures given; however, note that the bottom dimension for the can is shown as a diameter and must be halved before inserting into the formula:

$$\pi r^2 d = 3 \times 4cm^2 \times 15cm$$

equals 720cm³

DATA

At **Level 2** you will be expected to work with up to **20** sets of data displayed in various forms. You will need to understand, read, write, order, and compare data in tables, charts, diagrams and graphs.

Data can be displayed in not only numerical form, i.e. numbers, but also in various graphical modes. In other words, displayed in images, tables, charts, diagrams and graphs. This makes it possible to show comparisons, trends or even patterns in data.

Tables

A table is an easy way to record and display data. For instance, you have recorded the colour of cars in the college car park. There is a suitable title and column headings for colour and number of cars recorded. You may also place a total at the bottom to show how many cars there were in the car park.

Cars in the College	
Colour	**Number**
Silver	12
Blue	8
White	5
Black	10
Red	7

Tables can also display a vast amount of information and this may seem confusing at first glance, but learn to study carefully and understand the information displayed.

Here is an example of a typical bank repayment table. If you were interested in borrowing say £1,000 to buy a car, you can compare figures for **12** months, **24** months, **36** months, **48** months or **60** months.

Bank Repayment Period					
Amount of Loan £s	**12 Months**	**24 Months**	**36 Months**	**48 Months**	**60 Months**
£10,000	£916.67	£504.17	£369.72	£305.02	£268.42
£7,000	£641.67	£352.92	£258.81	£213.51	£187.89
£6,500	£595.83	£327.71	£240.32	£198.26	£174.47
£6,000	£550.00	£302.50	£221.83	£183.01	£161.05
£5,500	£504.17	£277.29	£203.35	£167.76	£147.63
£5,000	£458.33	£252.08	£184.86	£152.51	£134.21
£4,500	£412.50	£226.88	£166.38	£137.26	£120.79
£4,000	£366.67	£201.67	£147.89	£122.01	£107.37
£3,500	£320.83	£176.46	£129.40	£106.76	£93.95
£3,000	£275.00	£151.25	£110.92	£91.51	£80.53
£2,500	£229.17	£126.04	£92.43	£76.26	£67.10
£2,000	£183.33	£100.83	£73.94	£61.00	£53.68
£1,500	£137.50	£75.63	£55.46	£45.75	£40.26
£1,000	**£91.67**	**£50.42**	**£36.97**	**£30.50**	**£26.84**

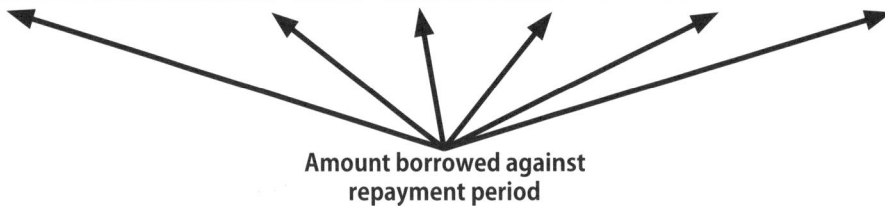

Amount borrowed against repayment period

Charts

Charts can be very appealing in appearance and can make viewing data easy, especially when comparing or contrasting. Rather than just showing numbers, they can be in the form of a **pie chart** or a **bar chart**, etc. Generally, at **Level 2**, the charts used tend to be either pie charts or bar (vertical, horizontal) charts. However, be careful that the charts you create make sense, have appropriate titles and axes and enhance the readers' understanding of the data you are presenting.

Pie charts

Here is an example of a pie chart showing the ages of people attending a sports club. There is an appropriate title and labels for the segments:

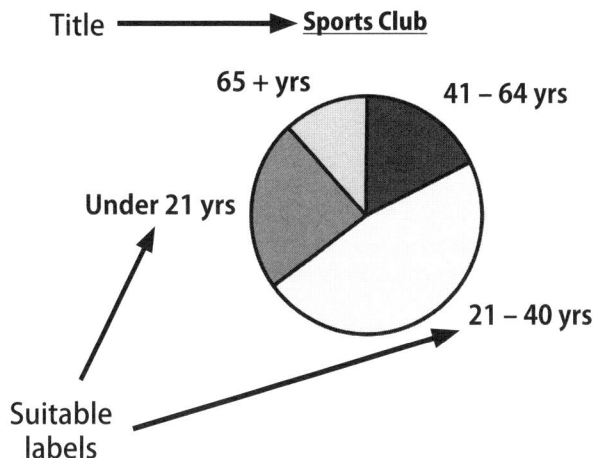

It can be clearly seen that the **most popular** age group is the **21 – 40 year olds**.

The **least popular** group is the **65 years and over**.

It is also easy to compare the age groups as a whole or individually.

However, the age groups can also have a numerical value applied of **figures or percentages and even degrees** displayed alongside the groups.

This adds more detail to the chart.

Bar charts

Bar charts can display single sets of data and also two or more sets of data. This is useful when **comparing information**. However, this is good as long as there are not too many bars displayed, as understanding the information can be confusing.

Bar charts tend to be displayed either **vertically** or **horizontally**. A title is generally included and axes displayed with suitable scales or labels.

Here is an example of a **vertical bar chart** of when people learnt to ride a bicycle:

Riding a bicycle

Number

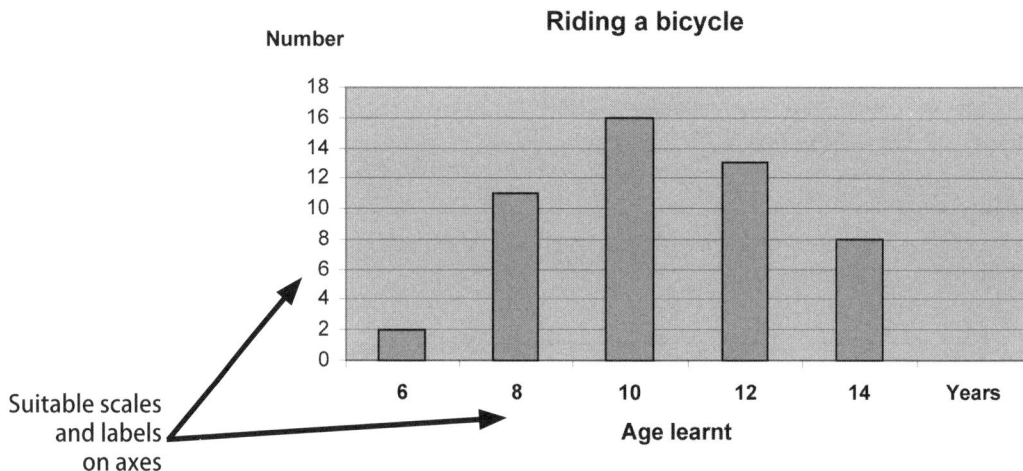

Suitable scales
and labels
on axes

Age learnt

Years

There is a title explaining the data (subject of the chart) and the **horizontal axis** shows the age learnt. The **vertical axis** shows the number of people who learnt. From the axes it can clearly be seen that **most** people learnt to ride a bicycle at **10 years of age** or that **only two** people learnt to ride a bicycle at **6 years of age**, etc.

Remember it is important to label the axes correctly, as failure to do so will lead to errors in data understanding.

Here is a **vertical bar chart** showing three sets of data over a three-month period. This is used when data needs to be compared. Importantly, there is a **key** identifying the separate bars on the chart.

Binary PCs

Key
Identifying the
chart bars

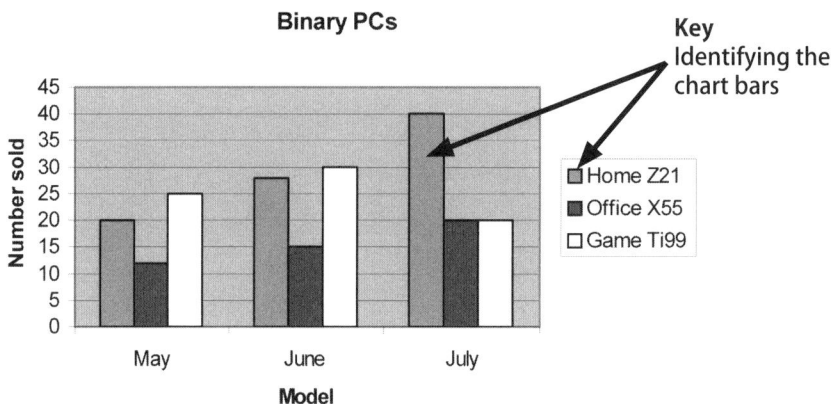

Number sold

Model

□ Home Z21
■ Office X55
□ Game Ti99

There is a title and suitable labels on the horizontal and vertical axes.

Q The following vertical bar chart is **missing** information. What is missing?

- International Airport -
Flights taking off late

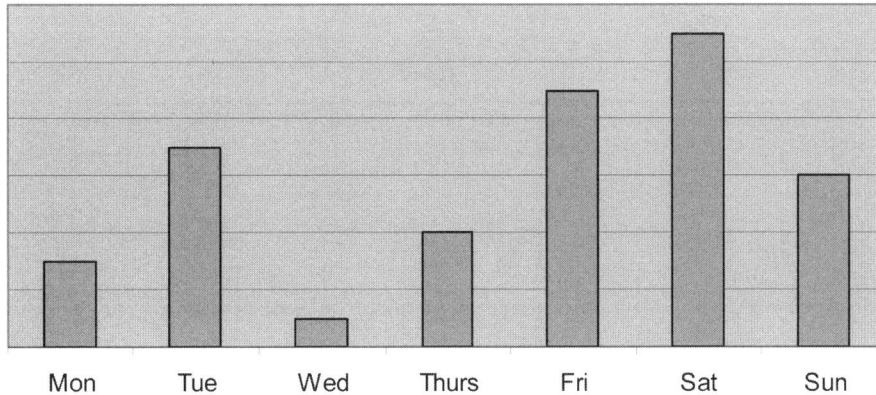

| Mon | Tue | Wed | Thurs | Fri | Sat | Sun |

✓ Looking at the bar chart the title looks correct as does the horizontal bottom axis. However, a scale and suitable label is **missing** from the **vertical** axis. So we don't know how many flights took off late.

Q From the horizontal bar chart how many people, **below** the age of **12**, learnt to ride a bicycle?

Age learnt to ride a bicycle

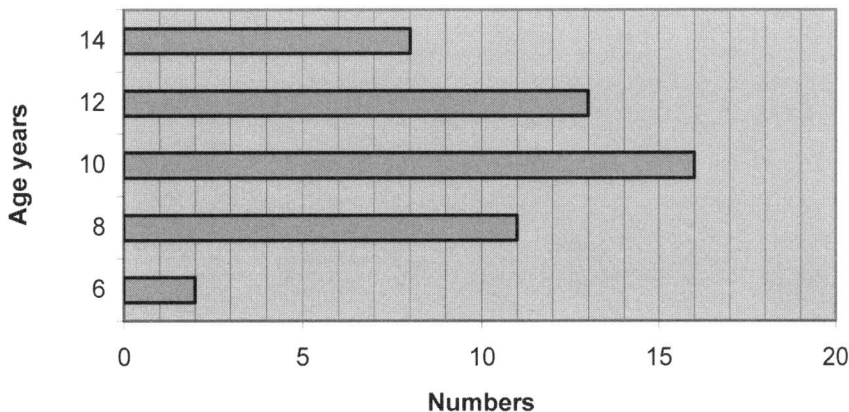

✓ We are only interested in the numbers **below 12** years of age. That will be the numbers for 10 years, 8 years and 6 years. They are **16+11+2 = 29**. We have **29** who learnt to ride a bicycle below 12 years of age.

DIAGRAMS

Diagrams are a good way of showing information such as the dimensions of a plan for an office. At **Level 2** you will be expected to work out missing dimensions and use these in your calculations of **area** or **perimeter**.

The following diagram has **missing dimensions**, but these can be worked out from other dimensions given, by either adding or subtracting values:

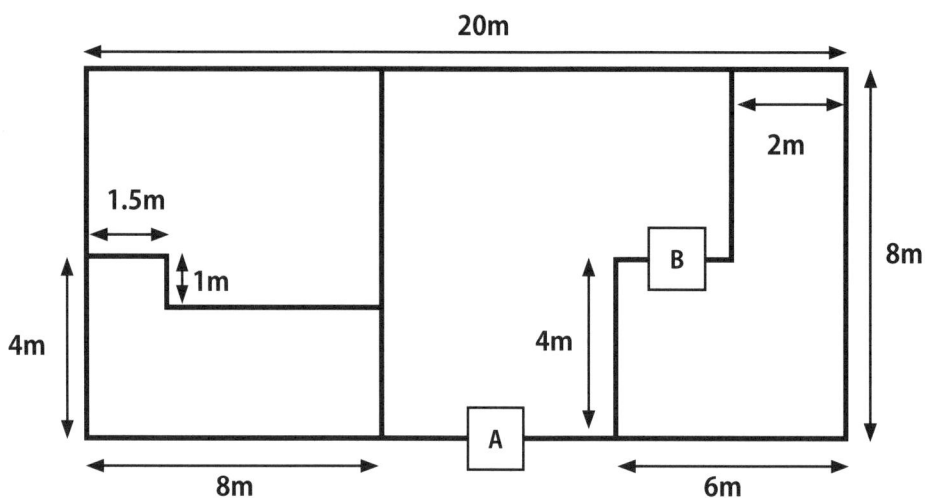

Dimension **A** can be found by adding the **8m** and **6m** either side together to get **14m** and then taking **14m** from **20m** which is the opposite side. So **A** equals **6m**.

B is found by taking **2m** from **6m**. Leaving **B** as **4m**.

Diagrams may also have other images inserted into them to aid the display of information or problems. The following is an example used in a volume problem:

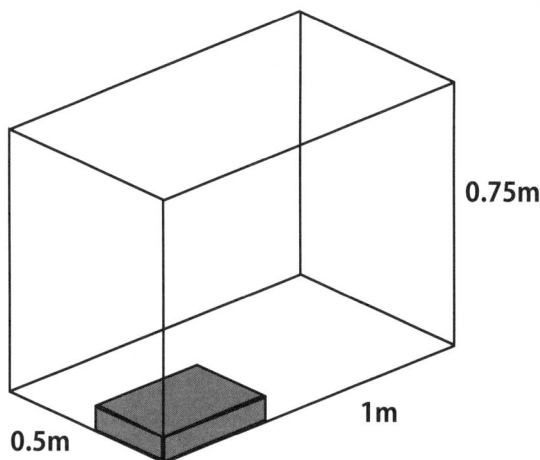

Q Using the following diagram, what are the missing dimensions?

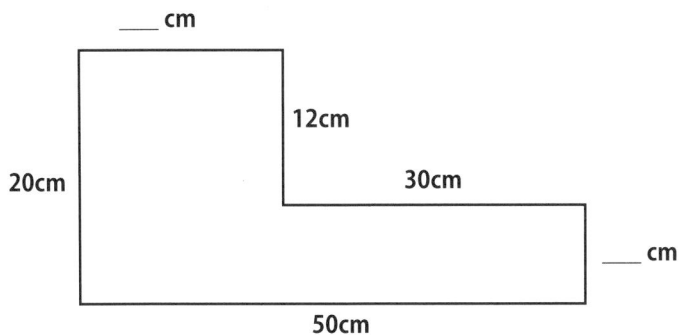

```
          ___ cm
     ┌──────────────┐
     │              │
     │              │ 12cm
20cm │              └──────────────────┐
     │                30cm             │
     │                                 │ ___ cm
     └─────────────────────────────────┘
              50cm
```

✓ Using the dimensions given, work out the **missing ones**. The top dimension is found by **50cm – 30cm = 20cm** and the side dimension is found by **20cm – 12cm = 8cm**. So the missing dimensions are **20cm** and **8cm**.

Graphs

Graphs are another useful way of displaying numerical information. Like bar charts they should have a **clear title** and **axes** using a **suitable scale**. Importantly, they must be labelled correctly. The following is a line graph displaying, the average temperature in degrees Celsius over five months in London:

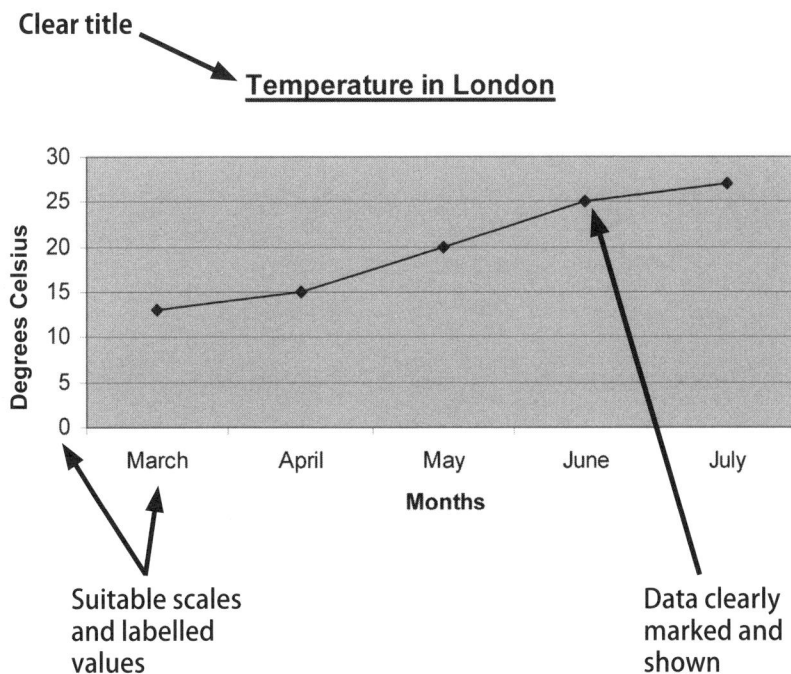

Clear title

Temperature in London

Suitable scales and labelled values

Data clearly marked and shown

The line graph is **clearly titled** and there are **suitable scales** for the horizontal and vertical axes. The axes are also **correctly labelled** showing **clear values**.

87

Two sets of data displayed

Customer Numbers

Key
identifying the
data lines

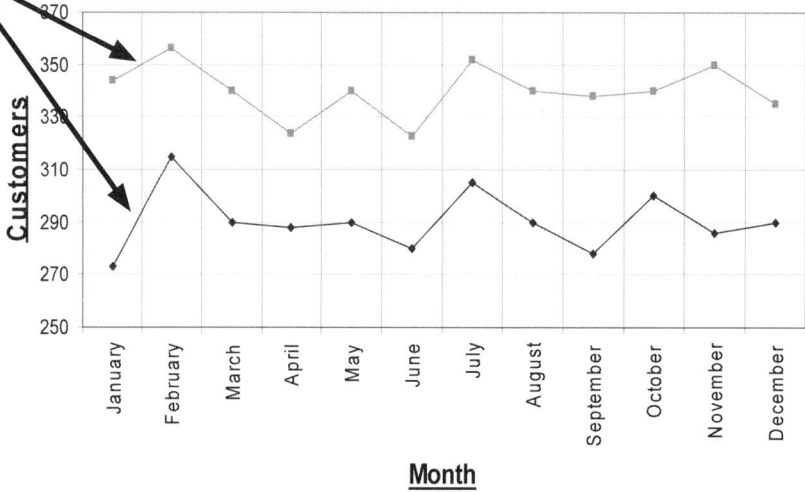

This is a **dual line graph** and it is used to display and compare **two sets of data**. There is a suitable title and the axes are clearly labelled; however, there is the addition of a **key** to **identify each line**. The top line on the graph is **A** because it has a **box** on the plot points and **B** has **diamonds** on its plot points. **Notice that in this example, the key is upside down to the graph.**

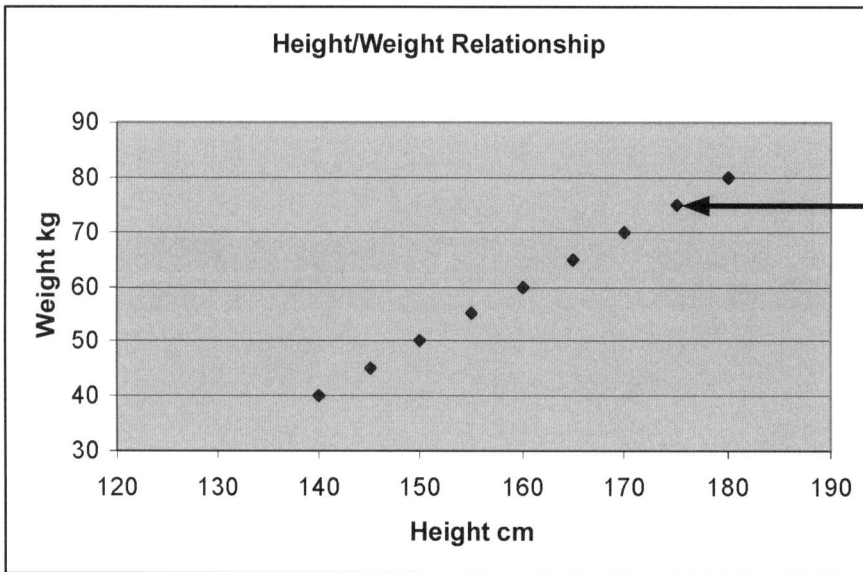

Plotted points showing a clear relationship between height and weight.

This is a **scatter graph**. Scatter graphs are used to show relationships against variables. In this case it can be noted that there is a clear relationship between a person's height and weight. In other words, the taller a person is, the heavier they tend to be. However this is not always the case.

Q From the table related to job types below, how many people have **nursing** jobs?

Job types	
Career	**Number**
Teacher	57
Doctor	15
Solicitor	12
Nurse	84
Engineer	57

From the table identify the career asked for **nursing** and read of the number of jobs. In this case there are **84 people with nursing jobs**.

Q From the following table, what is the gross profit in March?

Expenditure in £000s			
	January	February	March
Cash Sales	60.5	32.1	27.8
Callout Fees	1.5	2.8	3.1
Gross Profit	62	34.9	30.9
Expenditure			
Suppliers	20.3	10.2	0.9
Salaries	1.8	1.2	1.2
Bills	0.75	0.45	0.4
Totals	22.85	11.85	2.5

First of all it is important to note that the title states **expenditure in £000s**. This means that any figure in the table is actually hundreds of pounds. **So the gross profit for March is 30.9 £000s and this equals £30,900**.

Q From the line graph provided find how many **kilograms** there are in **55lb**?

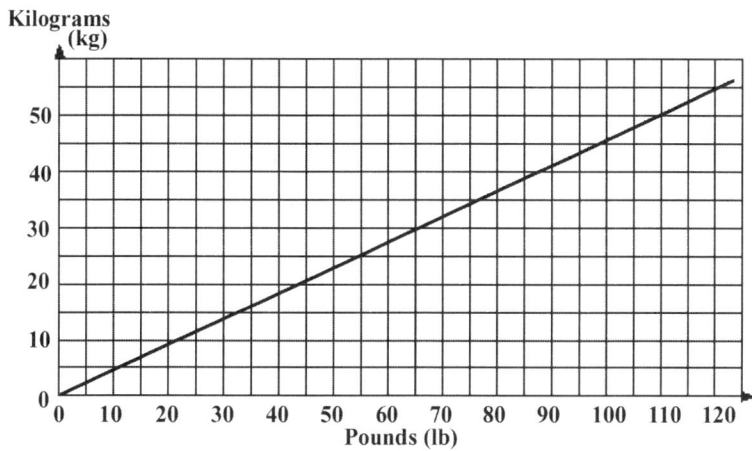

First, look along the **horizontal axis** as this shows **55lb**. Next, read up vertically to where **55lbs intersects** the diagonal line, then read off horizontally across on to the **vertical** kilogram scale. **This will show 25kg**.

Q Look at the dual line graph and identify the **least popular** month for **B** customers.

Customer Numbers

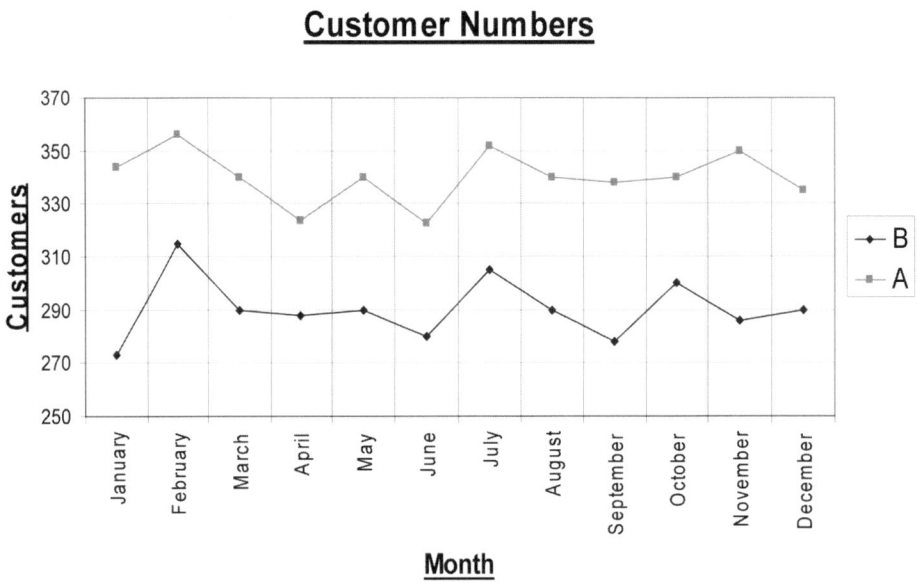

✓ The first thing to do here is **identify** which is **B** by using the key on the side of the line graph. **B has the diamond shape** on the plotted line. Then we look for the lowest point. **This is January.**

STATISTICAL MEASURES

Understand, compare – **mean**, **mode**, **median** and **range** in data in various forms.

Mean

Mean is another word for **average** and is treated in the same way. It can be used, for example, for finding the **average mark** in a set test or in **comparing** two sets of data.

If you have **ten** numbers and you wish to find the **mean**, first **add the ten numbers** together to **form a total**, and then **divide the total by ten** to find the mean number. **This works for any sets of numerical data.**

At **Level 2** you will be expected to work with at least **20** sets of numerical data.

Q What is the **mean** from the following test scores?

23	31	29	23	26
31	34	29	29	25

✓ Add the **ten** numbers together to form a total, **23+31+29+23+26+31+34+29+29+25 = 280**

Divide **280 ÷ 10 = 28** as the mean number.

Mode

Mode is the most frequently occurring data in a set of data. It is sometimes called **modal** and it can also be found in various forms of data, such as frequency tables or pictograms.

At **Level 2** you will be expected to work with at least **20** sets of numerical data.

Q What is the **modal activity** at the following sports club?

Sport activity	
Football	12
Boxing	6
Rugby	15
Swimming	8
Basketball	9

From the table of information it can clearly be seen that **rugby** is the mode.

Q What is the **modal age** of car plant employees?

Age of car plant employees	
Age Band	Number
20 – 29yrs	27
30 – 39yrs	72
40 – 49yrs	34
50 – 59yrs	25

Again it is easy to see that **30 – 39yrs** is the modal age.

Q Using the pictogram, what is the **modal** car **colour** and how many are there?

College car park		
Car colour		Number
silver	🚗🚗🚗🚗🚗🚗	12
blue	🚗🚗🚗🚗	8
white	🚗🚗🚗	5
black	🚗🚗🚗🚗🚗	10
red	🚗🚗🚗🚗	7
Key: Each 🚗 = 2 cars Each 🚙 = 1 car		

Silver is clearly the modal colour of car. The difficulty here is remembering to look for a **key** and then calculating the numbers from that. In this case a car image represents **2 cars**, so six images equals that figure **(6) multiplied by two = 12 cars**.

Median

Median is found by placing in order of size (**smallest to largest** or **largest to smallest**) a set of numerical data and then finding the middle figure. If it is between two sets of figures, then you must **add them together** and then **divide by two** to get the median figure.

At **Level 2** you will be expected to work with at least **20** sets of numerical data. Although this may seem time-consuming when placing the data out, it is necessary to get a correct and accurate result.

Q What is the **median time** from this set of times for running a 100 metres?

| 11.3 | 12.1 | 11.3 | 11.3 | 11.5 | 12.2 | 12.7 | 10.8 | 11.2 | 12.3 |

✅ Place in order of size **smallest to largest**.

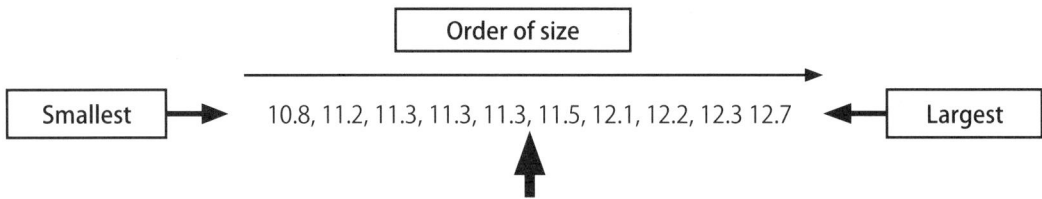

Order of size

Smallest → 10.8, 11.2, 11.3, 11.3, 11.3, 11.5, 12.1, 12.2, 12.3 12.7 ← Largest

The arrow shows the median figure between **11.3** and **11.5** – so adding **11.3 + 11.5 = 22.8** – then **22.8 ÷ 2 = 11.4** as the median figure.

Q A gardener records the temperature over ten days in November. What is the **median temperature**?

| 3 | 5 | -1 | -2 | -4 | -1 | 2 | 4 | -3 | -2 |

✅ Again place in order from **smallest to largest**. But this set of data has negative numbers.

-4, -3, -2, -2, -1, -1, 2, 3, 4, 5

The **arrow** shows the median figure between **-1** and **-1**. Therefore the median figure is **-1**.

Range

Range is used for finding the **spread** in sets of numerical data. It is calculated by finding the **largest figure** and taking away the **smallest (L – S)**. For example, if we were to look at a chart of temperatures and found **37°C** was the **largest** temperature and **7°C** the **smallest**, we would make the calculation as shown below:

37 minus 7 (37-7) = 30°C (**Range = 30°C**)

At **Level 2** you will be expected to work with at least **20** sets of numerical data.

Q A cyclist travels various distances over five days and records this information. What is the range?

Distance in kilometres
12.7, 17.9, 25.5, 11.2, 29.1

✅ **29.1** is the **largest. 11.2** is the **smallest**. So **29.1 minus 11.2 = 17.9 Km range**.

Q The weekly wage of some office workers was recorded; £345.21, £210.85, £177.25, £421.21, £367.91, £119.23, £112.23, £211.84, £201.48. What was the range?

✅ **Largest** £421.21 – **smallest** £112.23 = **£308.98 range**.

APPLICATION OF NUMBER: PART A, PRACTICE TASKS

TASK DESCRIPTION GRID

Number and title	Page	Activities	Refer to reference sheet(s) on page(s)
1 Binary PCs	95	Reading tables and calculating average and range. Converting information from tables into charts. Using percentages and formulae.	82 – 89, 90 – 92
		Converting exchange rates between currencies.	72, 80
2 Cyprus Nights	97	Reading tables to decide upon travel arrangements and calculate travel distances.	
		Converting between currencies.	
		Interpreting graphical and tabular information and describing what your results tell you.	81 – 89
3 Branching Out	101	Interpreting information and working with percentages, ranges and temperatures.	72, 92, 77
		Calculating volumes and reading scales.	80, 68
		Drawing a pie chart and describing what the results tell you.	83
4 Weld Tech	104	Interpreting graphical information and producing a report on a company's performance.	82 – 89
5 Plastered	106	Interpreting diagrams and calculating area. Using formulae and ratio. Reading scales to a given level of accuracy.	86, 78, 80, 67, 68, 66
6 The Scene	108	Interpreting tabular information and calculating median, mode and range.	90 – 92
		Calculating volume. Describing your results.	80
7 In Uniform	110	Interpreting maps to determine distances.	
		Calculating dimensions from drawings and plans.	86
		Converting between units of measurement.	74
8 Hairy Moments	112	Interpreting tabular information to calculate ratio and percentages.	82, 67 72
		Interpreting graphical information using angles and fractions and working to a given level of accuracy.	83 73, 66
9 Ice Cold	114	Interpreting graphical information to determine mean, median, range and mode.	90 – 92
		Using percentages and converting tabular information into charts.	72, 81-89
		Converting between currencies. Calculating dimensions from a diagram.	86
10 Apollo Cars	116	Calculating mean, median and range.	92
		Selecting ways to present information.	
		Using fractions. Interpreting diagrams.	69, 86

Number and title	Page	Activities	Refer to reference sheet(s) on page(s)
11 Tiny Steps	118	Interpreting scales and converting between units of measure.	68 74
		Reading graphical information and using formulae to convert units of measurement.	81 – 89, 80
		Selecting methods of presenting your findings.	
		Working to a given level of accuracy and using mean, median and range calculations in a report of your findings.	66, 90 – 92
12 Woodcraft	120	Interpreting diagrams and making calculations.	86
		Interpreting tabular information and using formulae.	82, 80
		Converting between units of measurement.	74
13 Making it Good	123	Calculating costs and profits.	
		Using formulae and volume.	80
14 Filly	125	Reading scales and converting between units of measurement and time.	68, 74 – 76
		Calculating mean, range and median.	90 – 92
		Calculating distances shown on a map.	
		Converting tabular information into graphical format.	80 – 89
15 Well Signed	128	Calculating area.	78
		Calculating ratio and converting information from tabular to graphical format.	67, 83 – 89
		Describing your results.	

TASK 1: BINARY PCs

Student Information

In this task, you will be reading tables, calculating averages, using percentages and formulae and converting currencies.

You will be converting information from tables into charts and describing the results.

REMEMBER:

Break the task down into small manageable parts and show your methods of working out and that you are checking your calculations for accuracy.

Describe what the information and results tell you.

Reading tables and converting information

Scenario

You are working for a multi-national computer company called **Binary PCs** and dealing with some of the numerical problems that you may come across within this employment.

Activities

You are looking at the pricing of computer components prior to them being sold.

The following is a table of computer peripherals that Binary PCs sells to the public.

Part	Item Description	Price £	VAT	Total cost	Selling Price £
Hard drives	IDE 20GB 5400RPM	35.00	6.13	41.13	49.35
	IDE 40GB 7200RPM	37.50	6.56	44.06	52.88
	IDE 40GB 7200RPM 8MB	45.00	7.88	52.88	63.45
	IDE 80GB 7200RPM 0/2MB	42.50	7.44	49.94	59.93
	IDE 80GB 7200RPM 8MB	45.00	7.88	52.88	63.45
	IDE 120GB 7200RPM	55.00	9.63	64.63	77.55
	IDE 160GB 7200RPM 2MB	60.00	10.50	70.50	84.60
	IDE 200GB 7200RPM 8MB	65.00	11.38	76.38	91.65
Monitors	TFT 15" SILVER/BLACK	120.00	21.00	141.00	169.20
	TFT 17" AVIDAV M1752S BLACK	160.00	28.00	188.00	225.60
	TFT 17" AVIDAV M1752S SILVER	155.00	27.13	182.13	218.55
	TFT 17" DIGIMATE SILVER	160.00	28.00	188.00	225.60
	TFT 17" GNR M173 MM	170.00	29.75	199.75	239.70
	TFT 17" NEOVO BLACK S-17A	225.00	39.38	264.38	317.25
	TFT 17" RELYSIS TL766 MULTIMEDIA	180.00	31.50	211.50	253.80
	TFT 17" TATUNG MM	160.00	28.00	188.00	225.60
Video Cards	AGP 128MB GEEFORCE4 TITANIUM4200	70.00	12.25	82.25	98.70
	AGP 128MB GEEFORCE4 TITANIUM4600	70.00	12.25	82.25	98.70
	AGP 256MB GEEFORCE FX5500	55.00	9.63	64.63	77.55
	AGP 256MB GEEFORCE FX5700V	75.00	13.13	88.13	105.75

Your line manager has asked you to perform some statistical analysis for costing and auditing purposes.

1 From the computer parts table calculate the **mean** (average) selling price of the hard drives, additionally calculate the mean selling price of monitors and then the mean selling price of video cards.

Check your answers with a calculator and correct any errors.
Remember to round off and use an **accuracy** of two decimal places to allow for currency.

2 Calculate the **range** selling price of each of the following: 1) hard drives, 2) monitors and 3) video cards.

3 Which computer part has the greatest price range? Describe why you think this.

4 Binary PCs manufactures and sells full computer systems. They offer various models based on popular specifications. These include the Z21, X55 and Ti99. Over a three-month period you have recorded the sales of these systems.

	Home Z21	Office X55	Game Ti99
May	20	12	25
June	28	15	30
July	35	20	20
Price	£600	£550	£750

Draw a **bar chart** with suitable scales and labels from this table of information to show a comparison in sales between the various models.

From this graphical display, **describe** the sales trend of each model.

5 During December there were 1,575 Ti99 models sold, in January there were 1,050 Ti99 models sold. **Describe** why you think there would be this difference between the two months.

6 Express as a **percentage** the Ti99 January sales compared to December sales. Clearly show your methods. Check your answer with a calculator.

7 An **Entry A1 PC** normally priced at £500 has been reduced by **25%**. What would the new selling price for this PC be?

8 Binary PCs has decided to print a full page advertisement in a popular computer magazine to improve sales.

Look at the advertising costs involved.

Dynamic Media Printing	
Initial advertisement page set up fee	£2,000
First six months advertisements	£1,000 per month
Additional advertisements placed for seven or more months	£500 per month

Show a formula that will calculate the cost of a 12-month series of advertisements.

9 A number of computer systems are sold in France. If the exchange rate is £1 = 1.5€ what would be the price in euros for the **Entry A1 PC** at £500 and the **Game Ti99** at £750?

10 A computer shop in France imports computers from Binary PCs. They import **10 Entry A1 PCs** and **10 Game Ti99 PCs**.

There is a shipping cost of £2.50 per kg.

If one computer unit weighs 4kg what is the total weight and total shipping cost of the 20 units in pounds sterling?

TASK 2: CYPRUS NIGHTS

Student Information	REMEMBER:
In this task, you will be using timetables, calculating travel distances and converting between currencies to calculate travelling costs. You will also interpret graphical and tabular information.	Break the task down into small manageable parts and show your methods of working out. Work to given levels of accuracy and describe what the information and results tell you.

Interpreting tabular and graphical information

Scenario

You are working for **Cyprus Nights**, which is a company building villas on the Island of Cyprus. You are to travel to Cyprus and work there.

Activities

1   The company that employs you has decided that you are to travel to London by train. Use the following train timetable to find the latest train from Durham that you can take to arrive in London by **10:30**.

Timetable					
Newcastle	05:40	06:15	07:05	08:10	09:30
Durham	05:50	06:35	07:25	08:35	09:55
Darlington	06:20	07:05	08:00	09:00	10:30
York	07:05	07:50	08:45	09:55	11:25
Peterborough	07:40	08:25	09:35	10:50	12:20
London	08:15	09:00	10:15	11:45	13:15

2   Find which train leaving Newcastle takes the **shortest** time to travel to London. Show the time in hours and minutes. Clearly show your methods.

3   You are to board a plane.

The flight from the UK will take 4½ hrs and Cyprus is **two hours** ahead of the UK in time. Calculate the time you will arrive in Cyprus if your flight leaves at 10:45.

4   When you landed at Paphos International Airport you travelled **89km** to Agros for the first construction job. Now this is complete you are travelling on to Nicosia for the next job. Using the distance table provided on *page 98*, work out the distance in kilometres from Agros to Nicosia.

	Polis	Agros	Pano Platres	Troodos	Paralimni	Agia Napa	Paphos Airport	Paphos	Limassol	Larnaka Airport	Larnaka	Nicosia	
C Y P R U S													Nicosia
												44	Larnaka
											4	48	Larnaka Airport
										70	66	81	Limassol
									66	136	132	147	Paphos
								13	56	136	122	137	Paphos Airport
							162	172	103	44	40	75	Agia Napa
						6	162	172	106	47	43	75	Paralimni
					152	152	46	56	42	116	112	69	Troodos
				5	141	141	44	54	35	105	101	74	Pano Platres
			24	18	139	139	89	99	33	103	99	53	Agros
		131	82	87	204	204	45	32	98	168	164	179	Polis
	Polis	Agros	Pano Platres	Troodos	Paralimni	Agia Napa	Paphos Airport	Paphos	Limassol	Larnaka Airport	Larnaka	Nicosia	**Distance in km**

5   When that job is completed there is a new job to travel to in Agia Napa. What is the **distance** from Nicosia to the new job?

6   Fuel costs 45 cents per litre in Cyprus and 100 cents equals 1 Cypriot pound. Calculate how **much** it will cost to travel from Agia Napa to Nicosia if the vehicle travels 20km per litre of fuel. Clearly show your chosen method and round off to the nearest cent.

7   The following pictogram shows the number of holiday villa projects completed in Cyprus.

July	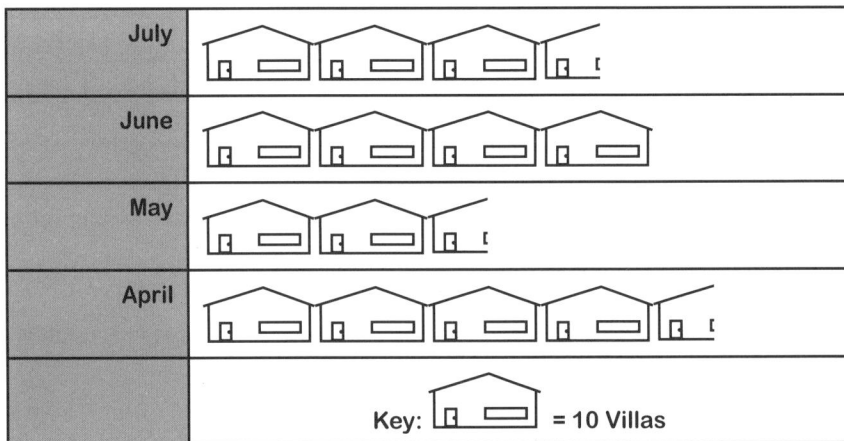
June	
May	
April	
	Key: = 10 Villas

How many villa projects were completed in the month of April, and how many projects were completed in the four months displayed?

8   The construction company has had a very successful year and you are to receive a production bonus of £1,976.

There are 180 people working in Cyprus and only 25% of the employees receive this bonus. Calculate how **many** people receive the production bonus.

9   From the 180 employees there are 72 people over 50 years of age. Express this as a **fraction** of the total.

10   From the table below based on hours worked in February, **round off** the times to the nearest hour.

	Day	Monday	Tuesday	Wednesday	Thursday	Friday	Saturday	Sunday
February	Week 1	8.5	9.25	11.0	12.75	8.5	6.25	6.0
Hours	Week 2	9.25	10.25	8.0	9.75	4.5	6.5	4.0
	Week 3	8.25	9.5	8.5	4.75	9.75	8.0	4.75
Nearest 0.25	Week 4	9.5	7.75	4.25	9.5	8.25	5.25	5.5

11   A driver is paid an hourly rate of £17.50 and a daily bonus of £25. Show a calculation that would give the correct answer for week 3 from the following monthly table. Check the answer with a calculator.

	Day	Monday	Tuesday	Wednesday	Thursday	Friday	Saturday	Sunday
February	Week 1	6	9	11	12	8		
	Week 2	9	10	8	9	4		
hours	Week 3	8	9	8	4	9		
worked	Week 4	9	7	4	9	8		

12   Accommodation costs have been reduced by £175.25 per month from an original cost of £990. What is the new **cost** of accommodation?

13   Look at the vertical bar chart displaying Cyprus temperatures in August:

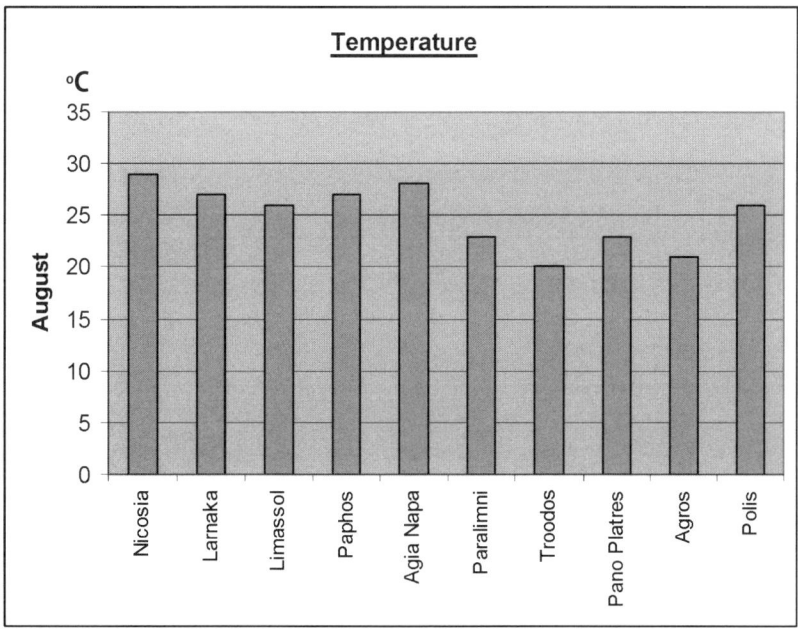

Which place has the **warmest** temperature, and by how many degrees? What is the **range** of temperatures shown?

14   The construction company is examining the differences in completed projects between the years 2004 and 2005. Calculate the difference in completed projects between these two years from the horizontal bar chart.

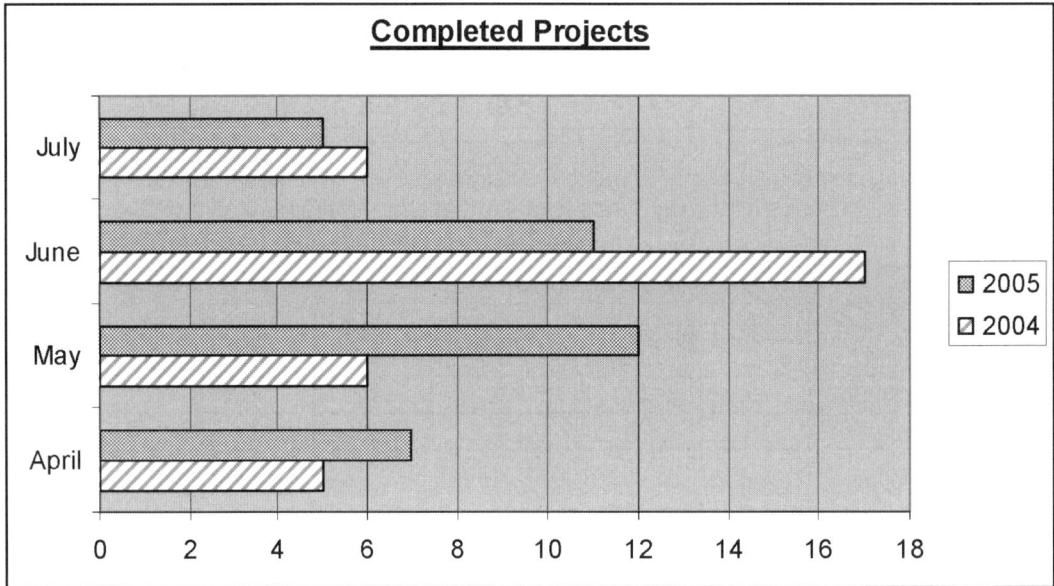

Completed Projects

15   **Describe** why there may be a difference between the two years shown.

TASK 3: BRANCHING OUT

Student Information

In this task, you will be interpreting tabular information and working with percentages, ranges and temperatures.

You will be reading scales and interpreting diagrams, creating a pie chart and describing what the chart tells you.

REMEMBER:

Break the task down into small manageable parts and show your methods of working out.

Check your methods of working, correcting errors and checking the result make sense and describe what the results tell you.

Presenting information and describing your results

Scenario

You are working as a tree surgeon for a company called **Branching Out**.

Activities

1   You are working in Canada with a group of tree surgeons conducting a survey on tree types. Part of this survey involves climbing trees for samples, but it is only allowed in certain temperatures because of health and safety reasons and therefore a daily record of the temperatures has to be kept.

Here is a table of recorded temperatures over a four-week period.

Week	Mon	Tues	Wed	Thurs	Fri	Sat	Sun
1	8	0	-2	5	4	5	1
2	8	0	3	6	-3	5	-1
3	7	6	0	-3	0	4	-3
4	8	7	1	2	-1	2	-4

From the table, what is the **fraction** of days where the temperature recordings were **below 0°C**?

2   Describe this fraction as a **percentage** from the total. Clearly show your methods.

3   What is the **range** of temperatures over the four weeks?

4   A generator's fuel tank holds 50 gallons of fuel. If there are **4.55** litres to one gallon, how many litres of fuel to the nearest full litre are in the tank when it is full?

5   The fuel tank on a sky lift crane holds 60 gallons of fuel when full. Look at the fuel gauges on *page 102* and estimate how much fuel the sky lift crane will use in 1 day. Also estimate how much fuel is used over 5 working days?

Clearly show your methods.

Start of day

End of day

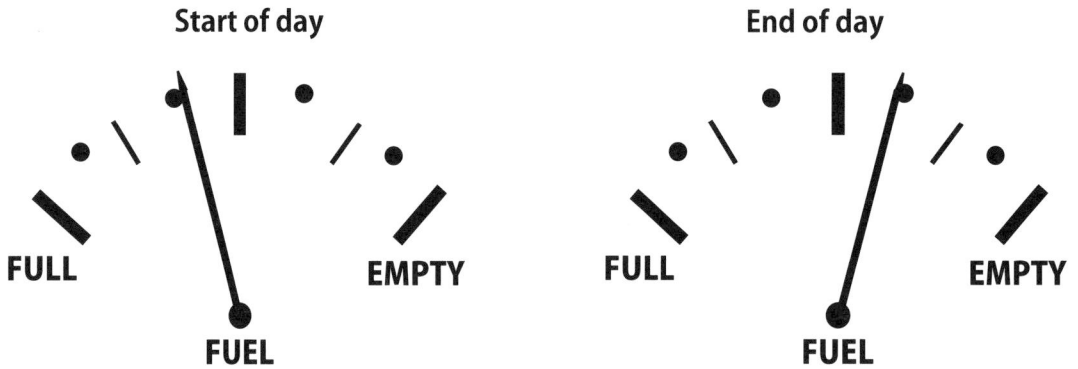

FULL   EMPTY   FULL   EMPTY

FUEL   **FUEL**

6   Look at this sketch and work out the length of safety rope required for the tree surgeon.

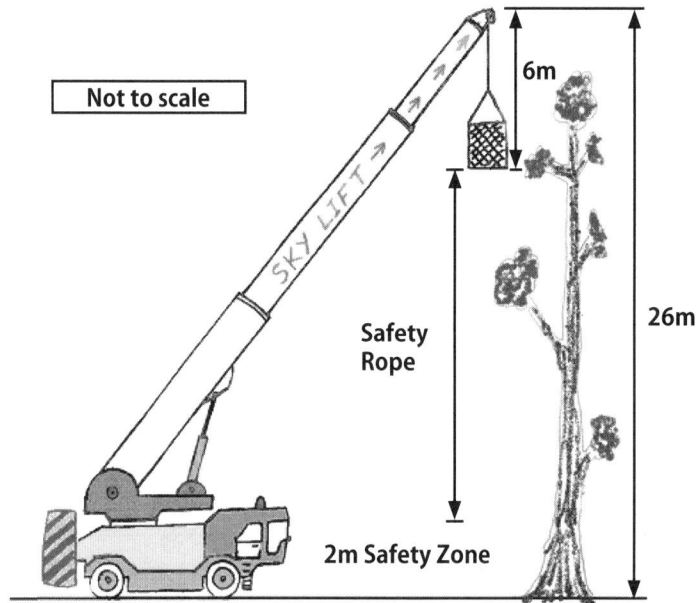

Not to scale

6m

SKY LIFT →

Safety
Rope

26m

2m Safety Zone

7   Oil is required for the sky lift crane on a regular basis. Which oil type gives the greatest range of temperatures required for working in Canada?

Oil Type	Temperature Range in °C
Titan Advance	-15 to 45
Synchro 2000	-20 to 55
New Oil Plus	-10 to 50
Oil Tech 65	-15 to 65
Super Lube 25 45	-25 to 45

8   Estimate the height of this bungalow if the tree is approximately 25m high.

9   During a recent survey of trees you have noted the following tree types in a particular area. This information was recorded in the following frequency table:

Tree Type	Oak	Douglas Fir	Sycamore	Beech
Frequency	35	24	42	19

Construct a pie chart using segments in degrees to display this information. Clearly show your calculations. Demonstrate an alternative way to calculate the degrees and check your answer with a calculator.

From the pie chart, write a brief summary of what your results tell you.

TASK 4: WELD TECH

Student Information

In this task, you have to interpret graphical information and produce a report.

You will select a suitable format in which to present information and justify your comments.

REMEMBER:

Break the task down into small manageable parts.

Show your methods of working out and describe your findings and methods.

Interpreting graphical information

Scenario

This task involves working as a Manager in an engineering company called **Weld Tech** and dealing with everyday numerical problems.

Activities

1   You are to produce a report on the business trading which compares and contrasts the trading for the years 2004 and 2005.

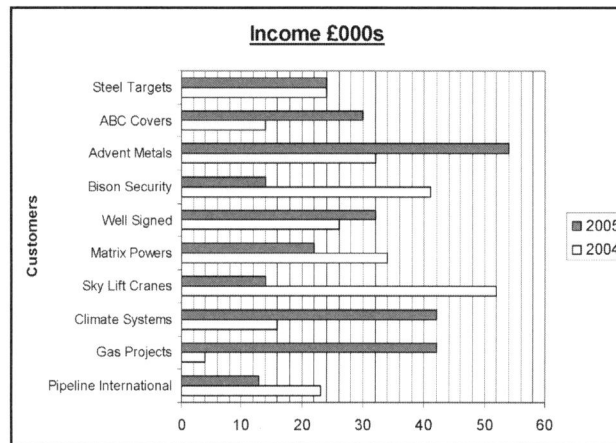

Income £000s

From this horizontal bar chart **identify** Weld Tech's best customer(s) in 2004 and 2005.

2   Which customer has increased turnover the **most** from 2004 to 2005?

Calculate this as a percentage **increase** and clearly show your methods.

3   Identify which customer has **decreased** turnover the most from 2004 to 2005 and express this as a financial figure. Show your methods.

4   Find the **total** income from the customers in 2004 and then in 2005 and work out the difference as a financial value.

5   Express the change between the years as a **percentage**, identifying either an increase or decrease in value.

Select and clearly show your **methods** and use a calculator to an accuracy of 1 decimal place.

6   Work out the **mean** income for 2004 and 2005 and also find the **range** for the two years.

7   **Group** the data from **Activities 5** and **6** into a suitable table or chart as part of a report.

        Describe the results and comment on your findings. Use the answers from previous tasks to validate your comments as necessary.

8   Look at the following engineering drawing for a component which is to be manufactured.

        What would be the life-size dimensions drawn on the following diagram for the link pin? Clearly show your methods.

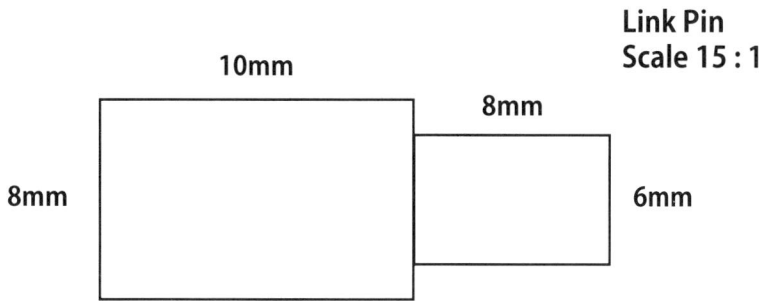

**Link Pin
Scale 15 : 1**

10mm

8mm

8mm

6mm

Diagram not to scale

TASK 5: PLASTERED

Student Information

In this task, you will be calculating area from diagrams and working with area calculations to determine dimensions and job costs.

You will use formulae and ratio and read scales to a given level of accuracy.

REMEMBER:

Break the task down into small manageable parts.

Show your methods of working out and that you are checking your calculations and answers for accuracy.

Area, ratio and scale

Scenario

You are working in the building industry as a plasterer. **Plastered** is a new business that employs plasterers to the house building trade. Today your job requires you to carry out a number of calculations.

Area calculations will help you with working out the times it takes to complete a job, and also help you to cost jobs based on how long the job will take to complete.

Activities

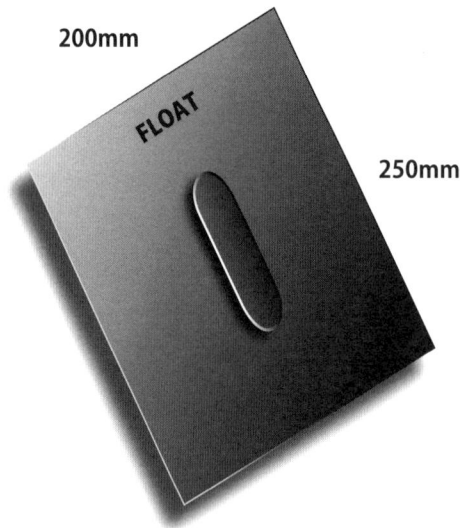

200mm

FLOAT

250mm

1   From the dimensions above, calculate the **area** of the plastering float in mm².

Calculate the **area** of the plastering float in cm².

2   Work out how many times the area of a float can go into a square metre. Clearly show your method.

3   You are required to plaster a ceiling and four walls after making allowances for fixtures.

Ceiling

4m

→ ● ←

1m Light fixture

8m

$\pi = 3$
$r = 0.5$

Calculate the area of the ceiling after allowing for the light fixture. Assume that the area of the light fixture is found by πr^2.

Room's Walls

1m

Door

2m

0.5m

2.5m

Window

1m

1m

8m

3m

1.5m

Fire

1m

3m

4m

4   Find the total area of the walls of the room after allowing for the door, window and fireplace.

5   Add the area of the ceiling **together** with the area of the room to form a **total** area. Use an accuracy of **two** decimal places.

Round off this answer to the nearest full metre.

6   To cover one square metre of wall with plaster generally costs £2.50 in materials. Calculate the **full** cost of plastering the complete room.

7   Assume it takes one hour to plaster 10m², **estimate** the time it will take to plaster the full room.

8 Plaster comes in large bags prior to being mixed with water.

Look at the scales (right) for one bag of plaster.

What is the reading on the scale to the **nearest** 250 grams?

9 Plaster is usually mixed in the ratio of weight by using 1.5 parts plaster to 1 part water (**1.5 : 1**). What weight of water for mixing would be required for 9kg of plaster?

Clearly show your method of calculation and include a check calculation.

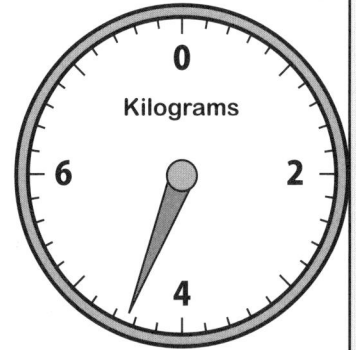

TASK 6: THE SCENE

Student Information	**REMEMBER:**
In this task, you will be interpreting tabular information to calculate the range of alcohol consumed and describe what the result tells you.	Break the task down into small manageable parts and show your methods of working out. Describe what your results tell you.

Interpreting tabular information

Scenario

You are employed as a Bar Manager in the student union bar at a popular university.

Today you will be involved with calculating alcohol consumption in the bar and working out bar profits.

Activities

1 As part of your role as a Manager in the bar, you record information on types and quantities of alcohol consumed.

Type	Sept	Oct	Nov	Dec	Jan	Feb	Mar	Apr	May	June
Lager pints	600	710	810	1210	510	650	700	820	610	860
Beer pints	420	580	640	710	220	340	490	540	430	510
Cider pints	340	440	490	510	180	290	360	480	400	490
Wine glasses	120	230	290	370	140	210	280	340	280	300
Alcopop bottles	460	580	620	1400	240	350	580	830	630	840
Spirit shots	90	160	290	410	130	210	280	290	240	250

From the information on alcohol consumption, **calculate** in which month the most alcohol was consumed.

Describe why that might be.

2   Which was the most **popular** drink type in April?

3   Find the **mean** number of glasses of wine and spirits sold over the ten-month period. What is the difference between the two mean figures?

4   Show the **median** figure of spirits consumed and clearly show your method.

5   Calculate the **range** of beer sold.

     What is the **modal** amount of spirits consumed?

6   Using the following table and the one on *page 108*, work out the total units of alcohol for cider consumed over the ten-month period.

Alcohol Type	Measure	Units of Alcohol
Beer, lager, cider	Pint	2
Wine	Glass	1
Alcopop	Bottle	1
Spirits	Shot	2

7   Beer is purchased at £0.85 per pint and sold at £2.05 per pint.

     Referring to the table on *page 108*, how much profit has the bar made in January?

8   Spirits come in 2 litre bottles and are sold in 25ml measures.

     How many measures can be obtained from one bottle?

9   If a 2 litre bottle of spirits costs £18 to purchase and a 25ml measure is sold at £1.25, what is the **total** made?

     What is the **profit** made by the bar? Clearly describe your chosen methods and why they meet your purpose.

TASK 7: IN UNIFORM

Student Information

In this task, you will be reading a map to determine distances.

You will calculate dimensions from a plan and convert between units of measurement.

REMEMBER:

Break the task down into small manageable parts.

Show your methods of working out.

Calculating dimensions

Scenario

These tasks are related to life in the Uniformed Services which cover a large area of employment, including the military and public services, together with working in the private sector, for example security companies.

Today your work involves a variety of army-related tasks.

Activities

1   You are to travel to an island for exercise manoeuvres. Look at the following map:

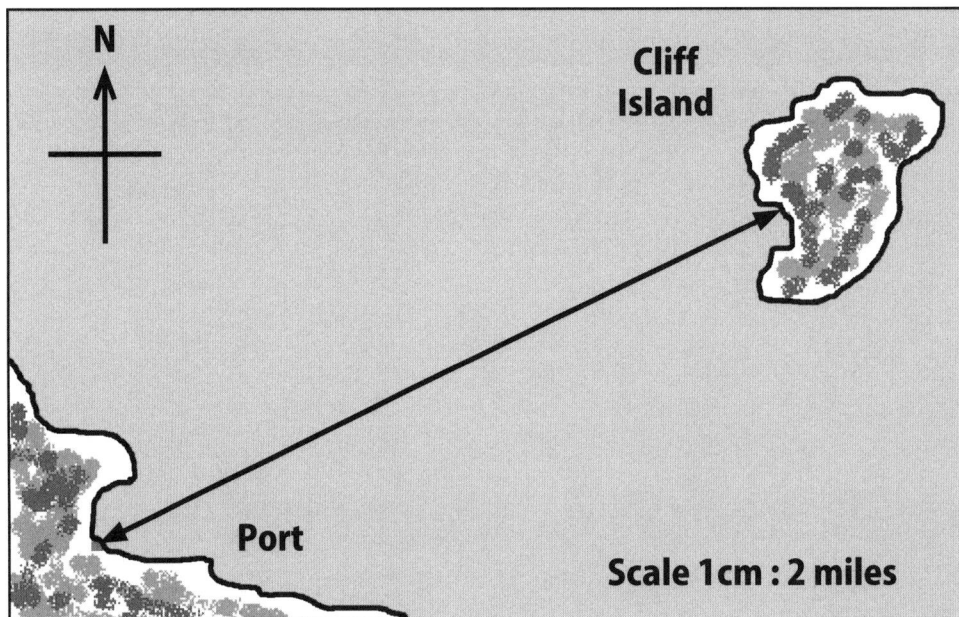

Using a ruler, measure the distance between the Port and Cliff Island to the nearest centimetre and **calculate** the distance in miles using the scale indicated.

2   Hood Island is north of Cliff Island (not shown on map) and is 32 kilometres wide. If 1.6km is approximately 1 mile, how many miles **wide** is Hood Island?

3   The scale drawing of this front line attack battle tank is **1 : 20**:

30cm

12cm

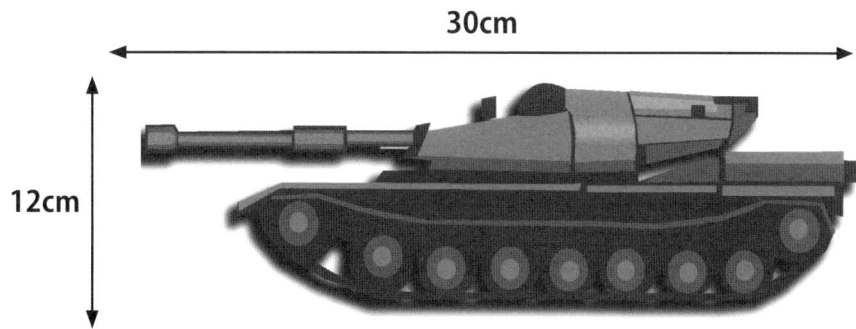

Calculate the **full** size dimensions in metres using an accuracy of one decimal place.

4   There is accommodation on the island. From the diagram of the barracks work out the **missing** dimensions for **A** and **B**.

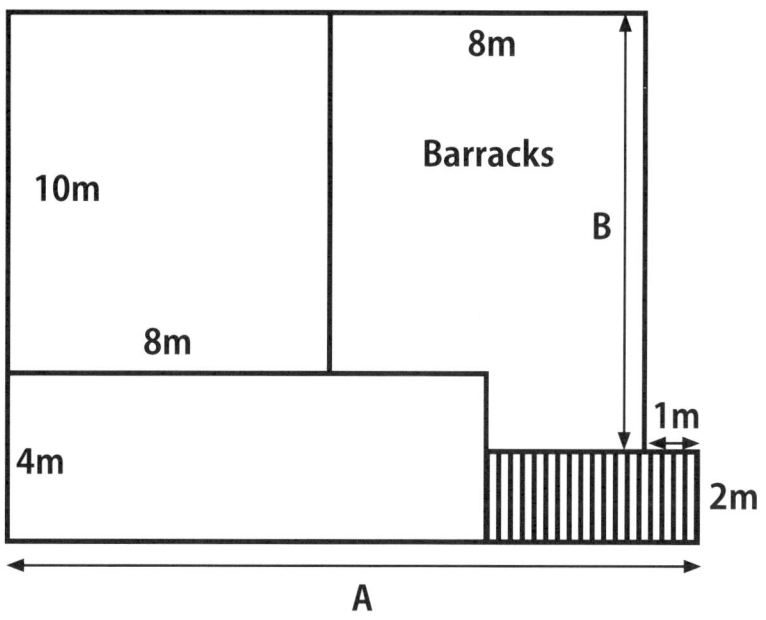

8m

Barracks

10m

B

8m

4m

1m

2m

A

5   A female soldier is 5 feet 6 inches tall.

1 foot = 12 inches and 1 inch = 2.5cm.
Show a calculation method to find the soldier's height in centimetres.

TASK 8: HAIRY MOMENTS

Student Information

In this task, you will interpret tabular information.

You will interpret graphical information using angles and fractions.

You will prepare graphical information to a given level of accuracy.

REMEMBER:

Break the task down into small manageable parts.

Show your methods of working out.

Compare sets of data and justify your findings.

Interpreting tables and preparing a chart

Scenario

The hairdressing industry is becoming more and more popular as an area of work for both females and males. You work in a salon called **Hairy Moments**.

Activities

1   The receptionist has recorded the number of appointments over a week:

Female	Monday	Tuesday	Wednesday	Thursday	Friday	Saturday
Morning	12	14	11	18	22	36
Afternoon	16	21	10	26	26	27
Evening	14	19	19	25	24	18
Home visit	3	4	2	5	8	

Male	Monday	Tuesday	Wednesday	Thursday	Friday	Saturday
Morning	6	4	9	8	10	18
Afternoon	13	11	12	15	18	12
Evening	17	16	15	19	26	6
Home visit		2	1		2	

    Form a **total** for both male and female and then calculate the **ratio** of females to males. Clearly show and describe your methods.

2   During a week in May there were 210 male customers having hair cuts. 4/5 of these also had another service. Express this as a **percentage** and the **number** of males.

3   25% of 380 female customers had more than two types of service. What is this as a figure and as a fraction?

4   On a busy Saturday there were 72 customers in total. Details of the customers' hair colour was placed in a pie chart.

How **many** customers are brunette?

5   What **angle** would hair colour red be on the pie chart?

6   What would the hair colour black be as a **fraction** of the total?

Hair Colour

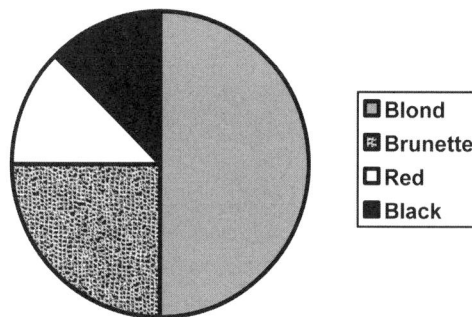

- ▢ Blond
- ▨ Brunette
- ▢ Red
- ▮ Black

7   Treatment types have been recorded over a three-day period. Draw a **pie chart** to show this information.

Cost	Treatment Type	Customer Numbers
£18	Cut 'n' blow	44
£25	Dye	28
£35	Extensions	58
£30	Highlights	14

Clearly describe you chosen method and correct any mistakes. Display the segments in **degrees**.

8   The salon is replacing its old stock with brand new equipment and your business needs to borrow £8,000.

Study the following bank repayment table and work out the **difference** in payments per month between 24 months and 48 months.

Use a check calculation to verify the difference.

Highlight the difference you found and describe why this may be. Justify your reasoning.

Bank Repayment Period					
Amount of Loan	12 Months	24 Months	36 Months	48 Months	60 Months
£10,000	£916.67	£504.17	£369.72	£305.02	£268.42
£7,000	£641.67	£352.92	£258.81	£213.51	£187.89
£6,500	£595.83	£327.71	£240.32	£198.26	£174.47
£6,000	£550.00	£302.50	£221.83	£183.01	£161.05
£5,500	£504.17	£277.29	£203.35	£167.76	£147.63
£5,000	£458.33	£252.08	£184.86	£152.51	£134.21
£4,500	£412.50	£226.88	£166.38	£137.26	£120.79
£4,000	£366.67	£201.67	£147.89	£122.01	£107.37
£3,500	£320.83	£176.46	£129.40	£106.76	£93.95
£3,000	£275.00	£151.25	£110.92	£91.51	£80.53
£2,500	£229.17	£126.04	£92.43	£76.26	£67.10
£2,000	£183.33	£100.83	£73.94	£61.00	£53.68
£1,500	£137.50	£75.63	£55.46	£45.75	£40.26
£1,000	£91.67	£50.42	£36.97	£30.50	£26.84

TASK 9: ICE COLD

Student Information	REMEMBER:
In this task, you will calculate mean, median, range and modal temperature calculations. You will calculate percentages and convert tabular information into graphical information.	Break the task down into small manageable parts and show your methods of working out. Clearly show and describe your methods used to check calculations. Verify your answers.

Statistical measurements, interpreting charts and diagrams and preparing a chart

Scenario

The food retail industry employs a great number of people. You are a supervisor working within the cold food section in a large retail outlet called **Ice Cold**. Today your work involves calculations related to the chill area and the cost of chilled food.

Activities

A large amount of food is stored within a chill area prior to being placed on display in the shop.

The temperatures are recorded and have been placed into the chart shown below.

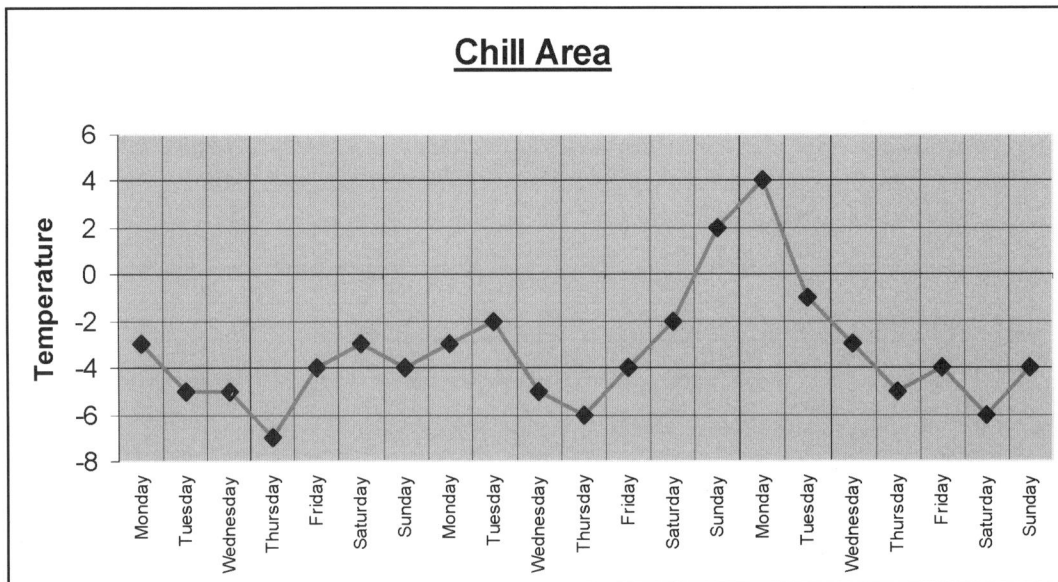

Chill Area

1   From the information illustrated in the chart calculate the **mean** temperature over the time period using 1 decimal place accuracy, and then find the **median** figure.

Clearly show and describe your methods and the choice of calculations used.

2   Show the **range** of temperature, and any **modal** temperature.

3   As a fraction, how many days was the temperature **below** –4?

4   For health and safety reasons, the maximum temperature allowed for the chill area is 2°C. This temperature was exceeded on one day.

As a result, the food stored at the time when the temperature rose will have to be discarded.

20% of the weekly stock of £84,000 was ruined as the result of the temperature rise.

Calculate this as a **figure**.

5   There were seven hundred items in the chill area; from this one hundred and seventy-four items were discarded.

Approximately what is this as a fraction?

6   Part of the discarded stock was fish, details of which are shown in the table below:

Fish Type	Cod	Haddock	Salmon	Plaice
Weight kg	28	20	14	10

The company plans to claim for the ruined stock from its insurance company.

Display this information in a **bar chart** with suitable scales and labels.

7   Cod is purchased from France at €2 per kg. How much does it cost for 28kg in £s sterling if the exchange rate is £1 = €1.60?

8   Your manager has requested some information about the chill area (including the Ice Blast). Calculate the area in metres squared.

Clearly show and describe your methods using check calculations to verify your answer.

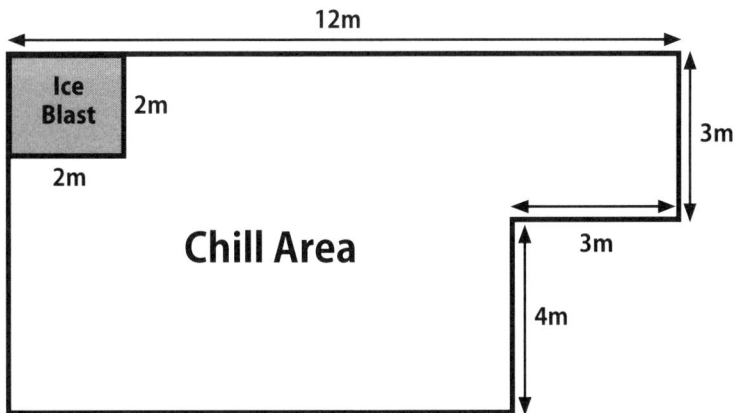

TASK 10: APOLLO CARS

Interpreting and presenting information

Scenario
The car industry is extremely competitive, with ever-growing numbers of cars on the road each year. Today you are working for a car sales company called **Apollo Cars**.

Activities
This is a sales table based on the cars currently in stock at a showroom in Edinburgh in the month of June.

Model	Price £	Engine size cc	Fuel	BHP	Turbo	Fuel Economy mpg	Insurance Group
V6 Panther	24000	2800	Petrol	160	Yes	28	18
Max D	14200	1600	Petrol	110	No	42	8
C12	10995	2000	Diesel	105	No	55	6
JJ 4Xt	16495	2500	Diesel	120	Yes	50	12
V12 ZZ	26750	3500	Petrol	160	No	25	20
Eco 2	8999	1300	Petrol	90	No	45	4
Eco 1	7495	1100	Petrol	75	No	50	3
Mitre xd	9200	1800	Diesel	90	Yes	50	5
Linus +6	7200	1600	Petrol	90	No	48	10
Swift 22	18250	2200	Petrol	125	Yes	34	16
Cat 3	22050	3000	Petrol	220	No	20	21
Neon Zx	24090	3200	Petrol	280	Yes	16	25
Radical d	16999	2000	Diesel	100	No	54	16
Luxor td	12045	1600	Diesel	110	Yes	58	16
Minx s	7495	1300	Petrol	90	No	54	5
Eco plus	5999	900	Petrol	75	No	56	2
Voyager 9	24999	2800	Diesel	140	Yes	28	12
Sun Tx	6599	1400	Petrol	90	No	48	18
Mars Vr	7999	1600	Petrol	95	No	44	16
Titan SS	11590	2000	Petrol	130	Yes	35	20

1. Work out the **total** stock value of the cars and the **range**, for insurance purposes. You can use a calculator to check your answers.

2. If the value of car stock exceeds a quarter of a million pounds, but is less than half a million pounds, there is a **25%** increase on the standard yearly insurance policy.

   If the policy normally costs £6,700, how **much** is the new policy? Show your chosen methods clearly and describe how you got your answer.

3. From the selling price of the vehicles, work out the **mean** value and also the **median** figure. Select a way to display these results and use an accuracy of two decimal places for currency.

4. **Group** the car data into a suitable table using price bands of £5,000 and model.

5. As a **fraction**, how many cars exceed a price value of £20,000 and what is this **total** value?

6. Calculate the **area** of the Car Showroom, Parts Reception and the Garage Workshop separately in m², and then form a **total** area of the premises.

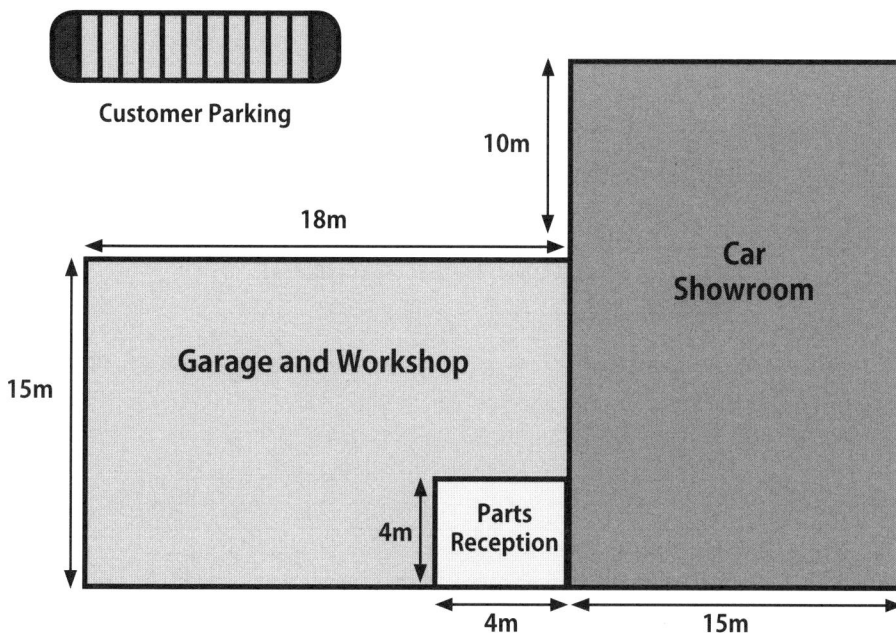

Customer Parking

Garage and Workshop

Car Showroom

Parts Reception

18m · 10m · 15m · 4m · 4m · 15m

7. Rental costs are £8.20 per square metre per year. Work out the rental for one year only, and then for five years.

TASK 11: TINY STEPS

Student Information	**REMEMBER:**
In this task, you will interpret scales and convert between units of measure. You will present your results and report on your findings.	Break the task down into small manageable parts. Show your methods of working out and work to given levels of accuracy. Select appropriate methods to present your information.

Scales and units of measure

Scenario

Childcare and Nursery staff need to be suitably qualified before being employed by a childcare organisation. Today you have to deal with the tasks as an employee of **Tiny Steps** Nursery.

Activities

1   Part of your job involves weighing children each week.

   Look at the scale and find the weight of a child in **pounds**. Record this weight.

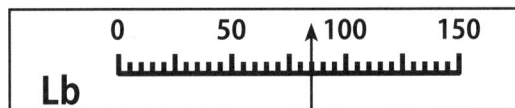

2   There are 14 pounds in one stone, change the weight to record in **stones** and **pounds**.

   Describe and clearly show your **method** of calculation.

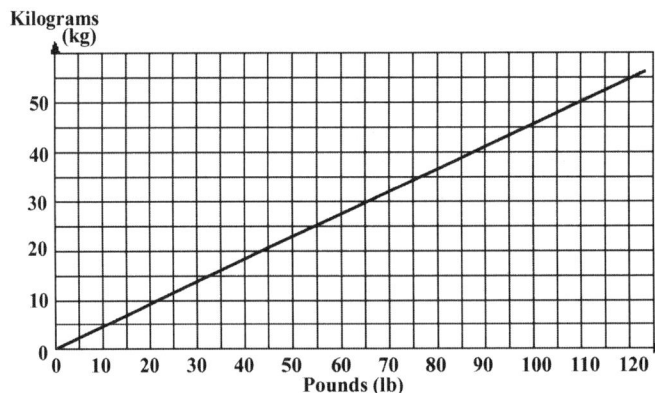

3   Using this line graph, **plot** the weight in pounds from **Activity 1** and then estimate the weight in **kilograms**. Record this information.

4   You are also required to monitor the temperature of the child.

   Read the scale of the thermometer and record the information.

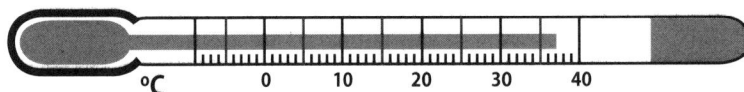

5   Using the following formula, convert the degrees Celsius into degrees Fahrenheit. Use an accuracy of one decimal place.

$$°F = (°C \times 9/5) + 32$$

Record this information.

6   Place the results from the previous recordings into a suitable **table** or **graph/chart** that includes headings:

Pounds; Stones and Pounds; Kilograms; Degrees Centigrade; Degrees Fahrenheit.

7   Look at the following information recorded over one week:

Group A					Group B				

Group A					**Group B**				
I. D.	Age	Weight kg	Height cm	Sex	I. D.	Age	Weight kg	Height cm	Sex
1	3yrs 6mths	28	91	M	1	2yrs 4mths	16	87	F
2	4yrs 2mths	32	105	M	2	4yrs 3mths	33	103	M
3	3yrs 2mths	29	93	F	3	2yrs 9mths	20	75	M
4	2yrs 9mths	24	74	M	4	2yrs 11mths	24	78	M
5	4yrs 8mths	40	108	F	5	3yrs 1mth	21	81	F
6	3yrs 6mths	31	87	F	6	4yrs 9mths	39	106	M
7	4yrs 7mths	34	102	F	7	4yrs 1mth	38	102	F
8	3yrs 11mths	27	101	M	8	3yrs 3mths	27	88	M
9	3yrs 3mths	22	96	M	9	4yrs 1mth	40	105	M
10	2yrs 11mths	18	78	F	10	2yrs 8mths	18	77	F
11	4yrs 1mth	34	99	F	11	3yrs 7mths	38	91	M
12	3yrs 5mths	25	86	F	12	2yrs 9mths	23	73	F
13	3yrs 7mths	27	94	M	13	3yrs 3mths	27	97	M
14	4yrs 1mth	31	98	F	14	4yrs 2mths	42	103	F
15	3yrs 4mths	26	82	F	15	2yrs 10mths	24	73	F
16	4yrs 7mths	33	108	M	16	4yrs 2mths	39	108	M
17	2yrs 11mths	19	74	M	17	3yrs 4mths	31	70	F
18	3yrs 3mths	24	82	M	18	3yrs 8mths	32	99	M
19	3yrs 6mths	29	84	F	19	3yrs 2mths	24	81	F
20	4yrs 2mths	31	104	M	20	3yrs 7mths	34	94	F

Calculate the **mean** and **mode** weight in kilograms for both groups.

Use an **accuracy** of one decimal place. Work out the **difference** between the two results and display this result.

8   Using the age column in **Group A**, find the **mean** age in months.

Repeat this for **Group B**.

Find the **median** age in months for **Group A** and for **Group B**.

9   Use the information on height to calculate the **range** of height for both groups.

Display the results from the two groups in a suitable way and **report** the main points of your findings.

TASK 12: WOODCRAFT

Student Information	**REMEMBER:**
In this task, you will interpret diagrams and tabular information to calculate costs, using formulae and working to a given level of accuracy.	Break the task down into small manageable parts. Show your methods of working out.

Area, diagrams and units of measurement

Scenario

Traditional construction trades, such as joinery, are becoming increasingly popular jobs.

You work for a construction company called **Woodcraft** and today have some numerical problems to solve for a customer who has asked for work to be done in his home and garden.

Activities

1   Woodcraft has asked you to undertake some jobs at a house. Look at the following plan view (shown from above) of the house and garden. A new wooden fence is required around the property.

        From the dimensions given, calculate the total **perimeter** of the garden to find the length of the fencing material required.

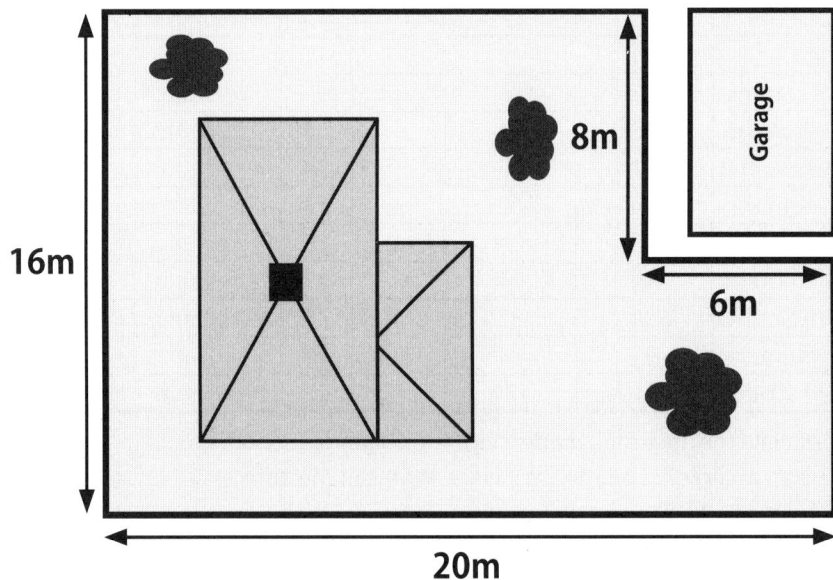

2   The customer has asked for a number of quotations for different types of fencing.

Look at the following and calculate the price for **each fence** type and select an effective way to show this.

Show all your calculations and **check** the answers with a calculator.

Woodcraft		
Fence type	**Height**	**Cost per metre**
Highland Lattice Panel	1m	£15.45
Dome Board Panel	1.5m	£21.99
Longwell Lap Panel	2m	£25.50
Vertical Spruce Panel	1m	£18.25
Crosshatch Pine Panel	1.5m	£20.99

3   The customer has decided to purchase the Dome Board Panel fence.

Calculate the **difference** in cost of the Dome Board Panel fence compared with the cheapest fence.

4   **VAT**, at **17.5%**, is to be added to the final bill of the Dome Board Panel fencing.

Using the following formula, work out the value of VAT and round off to two decimal places for currency. Show all of your calculations.

C is the cost of the fence

$$C \div 100 \times 17.5 = VAT$$

Now **add** the VAT to the final bill cost.

5   The entrance to the house needs a new replacement door and door frame. Look at the following diagram.

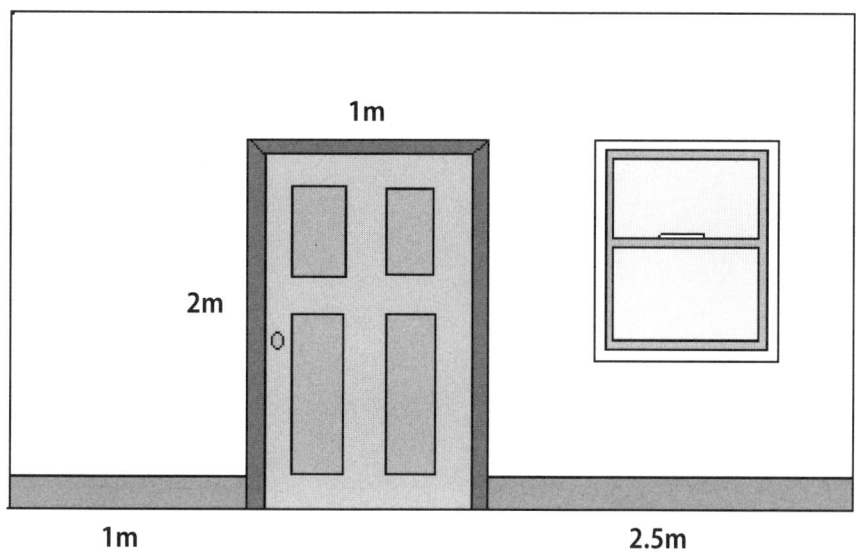

1m

2m

1m   2.5m

Diagram not to scale

If the door frame is 50mm wide, what size of door is required and what lengths of door framing? Give your answer in millimetres.

Convert this to show the dimensions in centimetres.

6   A roof requires replacement joists.

Using the following scale drawing, calculate the full length of the horizontal and angled joists in centimetres and then convert this to metres.

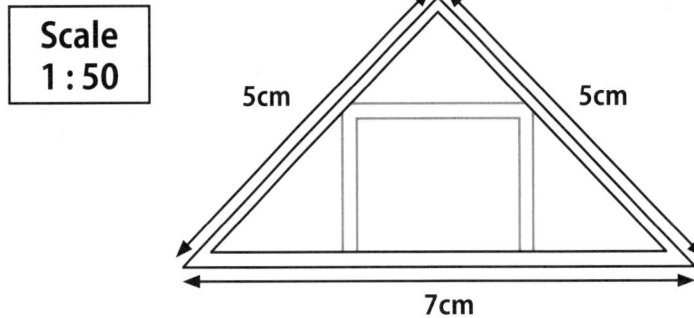

Scale
1 : 50

5cm   5cm

7cm

TASK 13: MAKING IT GOOD

Student Information	REMEMBER:
In this task, you will make calculations involving formulae and ratio, and work to a given level of accuracy. You will interpret diagrams and use calculations related to areas and costs.	Break the task down into small manageable parts. Show your methods of working out and that you are checking your answers for accuracy.

Volume and ratio

Scenario

You are working in the Catering and Hospitality industry and today are dealing with some of the problems related to a various organisations which make up that industry.

Activities

1   The ingredients for tea are purchased for a restaurant for a week in quantities of:

Item	Cost
Tea - 2kg	£6.00 per kg
Milk – 10 litres	£0.50 per litre
Sugar – 20kg	£1.50 per kg

Calculate the cost for the week?

2   Afternoon tea is served in a hotel restaurant at £0.65 per cup. If the ingredients make 200 cups of tea, how much profit will the restaurant make per cup to the nearest penny? Check your answer with a calculator.

3   Scones are made and served with the afternoon teas. Look at this recipe:

Ingredients for eight scones

200ml full fat milk

250g self-raising flour

150g sultanas

40g butter

Cook at 425°F for 20 minutes

How much butter is required to make 144 scones?

4   The oven uses a heat scale in degrees Fahrenheit. Use the following formula and calculator to convert 425°F to Celsius. Round off the answer to the nearest degree.

$$°C = (°F - 32) \times 5/9$$

5   Cream teas are served at a cost of £2.50 and during the afternoon 60 are sold, making a profit of £150. Show a check calculation to make sure the profit is correct.

6   During the week the customers using a restaurant were in the ratio of **3 : 7** males to females. If a total of 2,000 people came in, calculate how many males to females there were. Clearly show your method of working out.

In addition, the hotel business has a turnover of £12,500 in the week between the restaurant and accommodation in a ratio of **3 : 5**. Calculate the proportion of profit between the two.

7   New carpet is required for a hotel restaurant.

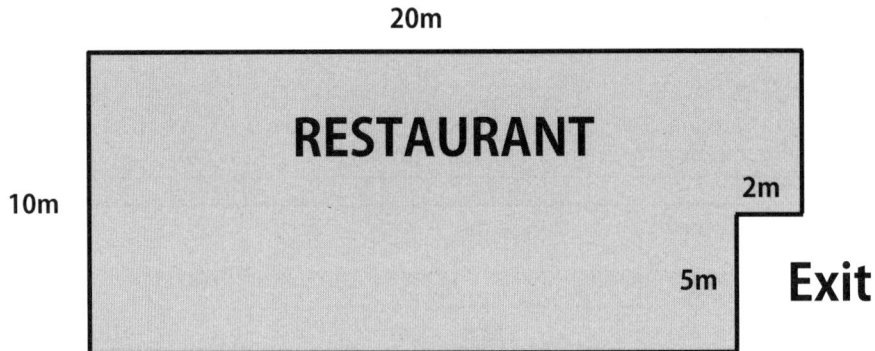

Calculate the total area in metres^2. Carpet costs £22.50m^2, work out the cost of the new carpet.

8   New carpet is also required at the exit. Look at the diagram below:

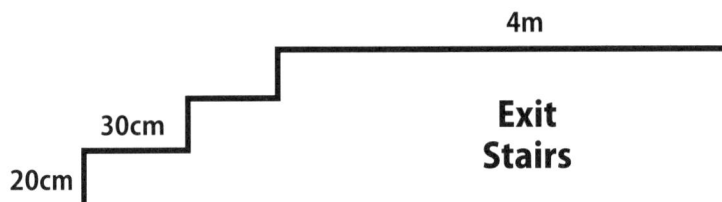

The exit stairs are 2 metres wide; work out the area from the sketch above and also the cost of carpet.

TASK 14: FILLY

Student Information

In this task, you will be reading scales and maps and converting between units of measurement and time.

You will convert tabular information into a pie chart.

REMEMBER:

Break the task down into small manageable parts.

Show your methods of working out.

Scales, measurement, time and statistical measures

Scenario

You are involved with the work of an Equestrian Centre called **Filly**. Today there are a variety of tasks on which you must work.

Activities

You have to work out how much straw to give "Brigadier", one of the horses.

1   Read the following scale and work out how many kilograms there are in one large bale of straw.

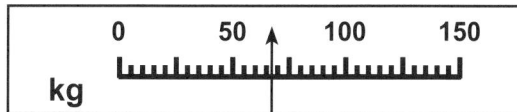

| 0 | 50 | 100 | 150 |

kg

2   How many kilograms are there in two bales of straw? In addition, calculate the weight in pounds if 1kg = 2.2lb?

3   If Brigadier weighs 218kg and needs to have 8kg per day, how much straw will this horse need in a week?

4   Look at the engine temperature gauge on a tractor. What reading does it show?

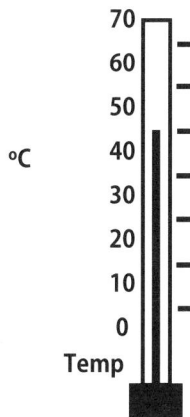

```
70
60
50
°C    40
30
20
10
 0
Temp
```

5   The tractor's maximum operating temperature should not **exceed** 70° for safety. How far **below** 70° is it in degrees?

6   The length of a wall has been measured with a metric tape measure. This measurement is shown on the diagram below. Using this length, calculate how many bales of straw, each measuring **2.25m** long, you can fit against the wall.

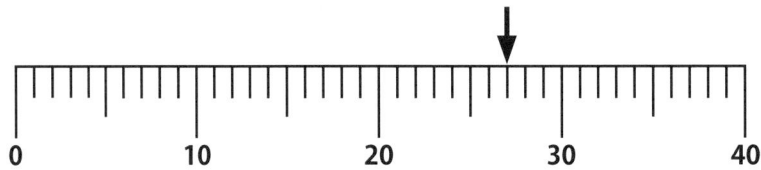

7   What measurement does the tape show, accurate to the **nearest** 10 metres?

8   A straw delivery for the horses is ordered for 11:55. Referring to the clock below, how much **longer** is there to wait in hours and minutes?

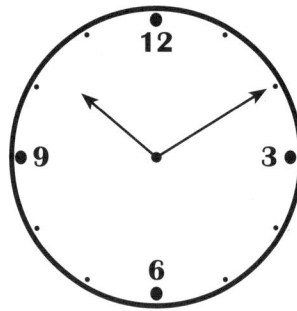

9   The next delivery is expected at the time shown on the following 24-hour clock. What time does it show using the 12-hour format?

13:25

10   Calculate the difference in time between the first straw delivery and the second delivery in the afternoon.

Show your answer in minutes, then show the time in hours and minutes.

11   In one week you are paid three hundred and forty-nine pounds and sixty-four pence. Write this figure out in **numbers**.

12   Your weekly wages are recorded in the following table. You wonder what your mean (average) wage is over 20-weeks.

Work out your **mean** wage using the table below.

Wage	£360	£240	£280	£320	£260	£220	£280	£260	£250	£280
	£345	£360	£305	£250	£270	£300	£270	£300	£240	£260

13   From the table, find the **range** of wages over the 20-week period and also the **median** wage.

Clearly show your working out and correct any errors.

14   During one week you travel to various equestrian centres. Look at the map on the following page and work out exactly how far in **miles** it is from Northcastle to Sothton?

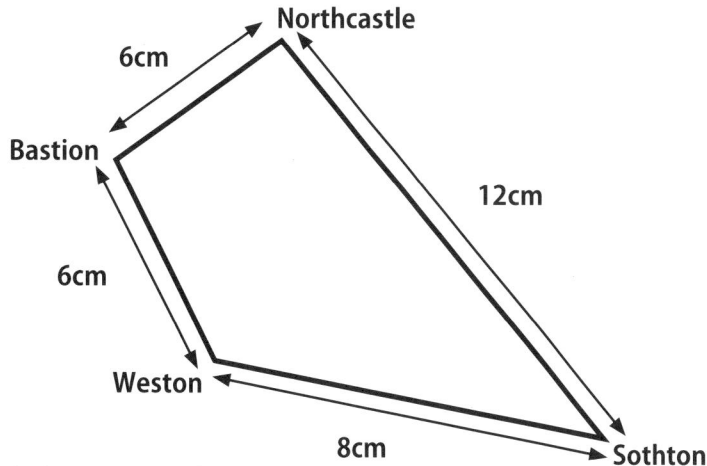

Scale = 2cm : 1 mile

15   Assuming you have travelled to all the places shown on the map, work out how far you have travelled in the full week.

16   These are the hours that you worked over the week:

Day	Monday	Tuesday	Wednesday	Thursday	Friday
Hours	6	10	8	4	8

Display these in a pie chart showing degrees. Clearly show your chosen methods.

TASK 15: WELL SIGNED

Student Information

In this task, you will be calculating area by interpreting diagrams.

You will calculate ratio and convert information shown in tabular format into a bar chart and pie chart, and then describe what the results mean.

REMEMBER:

Break the task down into small manageable parts.

Show your methods of working out and that you are checking for errors.

Describe the results.

Interpreting diagrams, using ratio and describing results

Scenario

You are working for a firm called **Well Signed**, which manufactures and installs hazard signs. Today's work involves calculating areas and using ratio.

Activities

As part of your job you are required to work out the area of each sign and the work area in which they are placed.

1 This **Danger** sign measures 80cm by 50cm and is to be installed in to a Sub-station.

Work out the sign area in cm^2.

2 This **Caution** sign measures 48cm by 32cm and will be placed in a workshop.

What is the area of the sign?

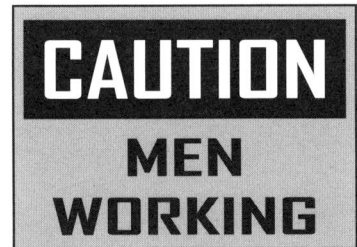

3 This **Danger** sign measures 1.2m by 0.8m and is situated in a transformer room.

Calculate the area of the sign.

Complete the following tasks on finding work area in metres squared.

To help you through these problems, break the task into smaller parts to find the solution.

4 Work out the **total** area of the Sub-station in metres squared.

Draw a scale diagram using 2cm to 1m of the Sub-station.

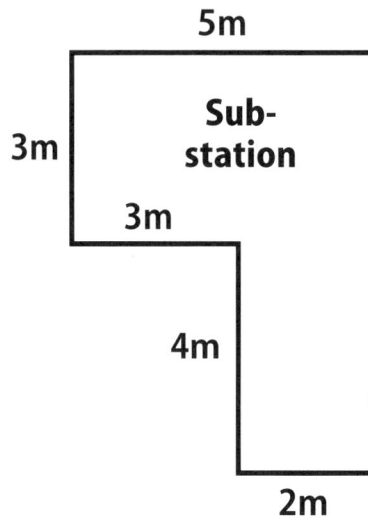

5 Find the **total** area in m² of the workshop by breaking the task into parts.

Draw a scale diagram using a scale of **2cm : 1m** of the Workshop.

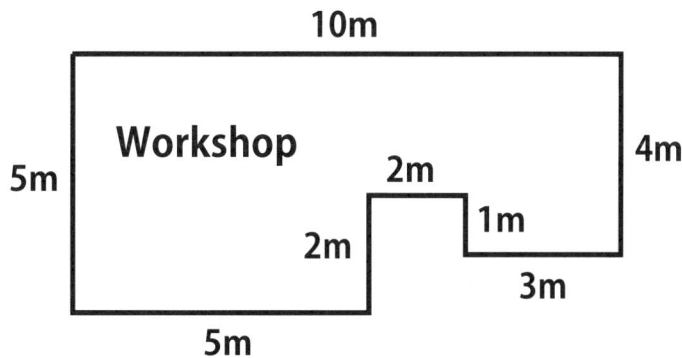

6   Calculate the **total** area of the Transformer Room below and clearly show your method.

Show an **alternative** way to find the total area of the Transformer Room.

Draw a scale diagram using a scale of **1cm : 1m** of the Transformer Room.

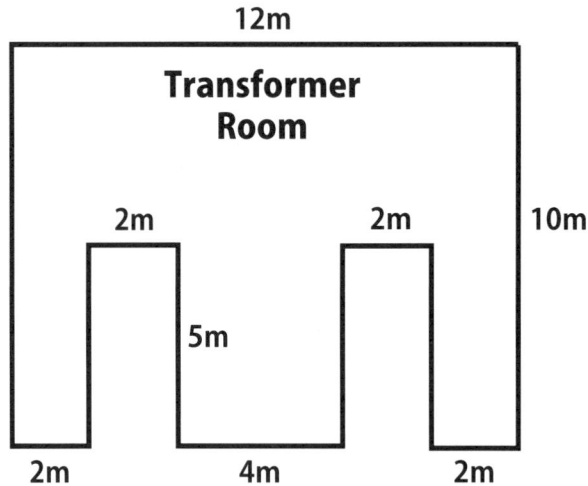

12m

Transformer Room

2m   2m   10m

5m

2m   4m   2m

7   Paint is mixed to create a shade of orange used on hazard signs. The painter uses three times as much red as yellow (**3 : 1**).

If she used 27 litres of red, how **many** litres of yellow paint will she use?

8   Certain types of signs are cemented into place. During the mixing process a chemical is added to the cement to act as a hardening agent.

It is mixed in a ratio of **3 : 5** chemical to water.

How **many** millilitres of each will you need to form 3 litres of mixture?

9   Paint is purchased in containers and used for the mixing of various paint colours.

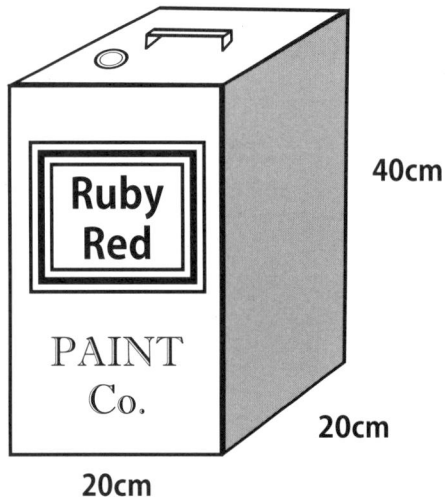

Ruby Red

PAINT
Co.

40cm

20cm

20cm

Ruby red paint comes in a standard size paint container.

From the dimensions work out the **volume** of the container in centimetres cubed.

2

10   Paint is then stored in large paint vats prior to mixing with other colours for painting signs.

Sky blue paint is stored in this vat. Look at the dimensions and calculate the **volume** of the paint in the vat in cm³.

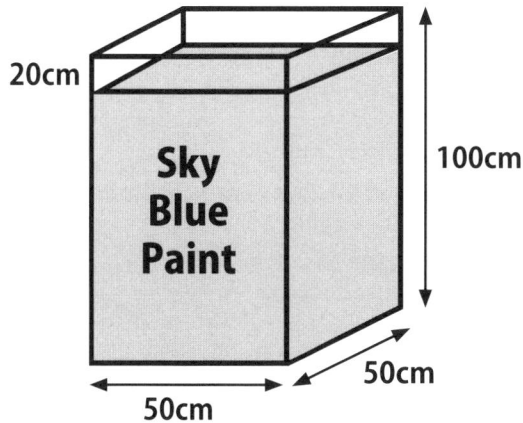

20cm

Sky
Blue
Paint

100cm

50cm

50cm

Note

The paint vat is **not** full.

11   After being manufactured, some hazard signs are packed into large containers for transporting.

However, the company is charged by the volume for the containers they use.

Look at the size of this transporting container and work out its volume in metres³.

Then **calculate** how many signs with a dimension of 79cm long by 49cm wide by 9cm deep can be placed flat inside the container.

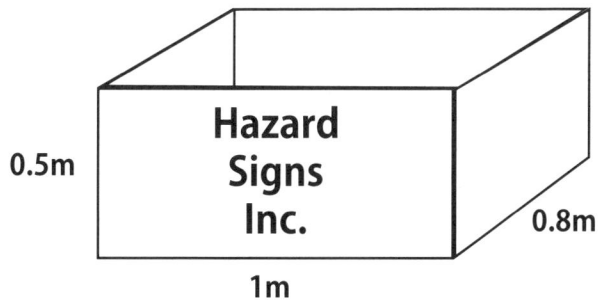

Hazard
Signs
Inc.

0.5m

0.8m

1m

12   During one week in June you record the number of signs produced. From the information shown in this frequency table, draw a suitable **vertical bar chart** using correct labels.

June week 1	Sign	Danger	Caution	Hazard	Beware
	Number painted	16	18	28	10

13   From the same table of information on signs produced, display the information as a pie chart.

Clearly show your working out and label the segments **correctly** with degrees. Describe why you think there are more signs of one type than others.

SAMPLE END ASSESSMENT

20 Multiple-choice questions

The following questions are multiple-choice. There is only one correct answer to each question.

Instructions

1   Choose whether you think the answer is A, B, C or D.

2   Ask your tutor for a copy of the answer grid (or download a copy from **www.lexden-publishing.co.uk/keyskills**).

3   Enter your answer on the marking grid at the end of the test.

4   Hand it to your tutor for marking.

An Application of Number Key Skills Level 2 External Assessment will consist of 40 questions and you will have **1 hour and 15 minutes** to complete them.

How will you select your answers?

If you are sitting your End Assessment in paper format – not doing an online test – you will have to select one lettered answer for each numbered question. The answer sheet will be set in a similar way to the example below:

1   [a] [b] [c] [d]

2   [a] [b] [c] [d]

Make your choice by putting a **horizontal line** through the letter you think corresponds with the correct answer.

Use a pencil so you can alter your answer if you wish and take an eraser to allow you to change your mind about a response. Use an **HB pencil**, which is easier to erase. (If you make two responses for any one question, the question will be electronically marked as **incorrect**.)

Take a **black pen** into the exam room because you will have to sign the answer sheet.

Your tutor has 100 sample End Assessment questions and you will be given these when your tutor considers you are ready to practise the questions.

QUESTIONS

1   A managing director's £15,000 car loses 30% value during a year. The HM Revenue and Customs will allow him to claim back **40%** of this loss.

Which calculation shows the **allowance** he can claim back?

A $\dfrac{15,000}{30} \times \dfrac{40}{100} = £200.00$

B $15,000 \times \dfrac{30}{100} + £4000 = £8,500$

C $15,000 \times \dfrac{30}{100} \times \dfrac{40}{100} = £1,800$

D $15,000 \times 100 \div 30 \div 40 = £1,250$

2   Khan buys six CDs on sale at £8.75 each. He pays the cashier £60. Which of the following methods will give him the closest **estimate** to his change?

A $60 - (9 \times 6)$

B $6 \times (9 - 60)$

C $60 + (6 \times 9)$

D $9 - (6 + 60)$

132

3 Times from a 100m sprint are recorded in the following table:

12.4	11.9	11.1	12.2	11.5
12.7	12.0	11.3	12.3	11.7
12.5	10.9	11.2	12.3	11.5
12.8	12.1	11.5	12.4	11.7

Which of the following is the **mean** time?

A   11.5

B   12.1

C   12.0

D   11.9

4 Using the 100m sprint table, what is the **modal** figure?

A   11.3

B   11.5

C   12.4

D   11.7

5 From the 100m sprint table, which of the following is the **median** time?

A   10.95

B   11.95

C   12.15

D   11.75

6 A gardener records the temperature in degrees Celsius in his greenhouse over a 28-day period of time:

Week 1	2	-1	-4	0	2	1	3
Week 2	3	-2	1	-3	0	1	-2
Week 3	5	3	2	7	2	-3	5
Week 4	1	2	3	-2	5	9	6

What is the **range** of temperature?

A   6°

B   8°

C   10°

D   13°

7 A manager of a computer shop has created a spreadsheet on business expenditure:

Expenditure in £000s			
	January	February	March
Cash Sales	60.5	32.1	27.8
Callout Fees	1.5	2.8	3.1
Gross Profit	62	34.9	30.9
Expenditure			
Suppliers	20.3	10.2	0.9
Salaries	1.8	1.2	1.2
Bills	0.75	0.45	0.4
Totals	22.85	11.85	2.5

From the spreadsheet, what level of **accuracy** has been used?

A   Nearest 10p

B   Nearest £1

C   Nearest £10

D   Nearest £100

8 An oil manufacturer produces different types of specialist oils for engines:

Oil Type	Temperature Range in °C
Synchro 2000	-20 to 55
New Oil Plus	-10 to 50
Oil Tech 65	-15 to 65
Super Lube 25 45	-25 to 45

From the oil supplier's brochure, which oil has the greatest temperature **range**?

A   Synchro 2000

B   New Oil Plus

C   Oil Tech 65

D   Super Lube 25 45

9   A pet shop records the length of guinea pigs on to a chart. The size is recorded in centimetres:

Pampered Pets Guinea Pigs				
12	21	16	12	19
14	17	22	15	20
17	21	18	14	10
9	15	17	11	16

What is the **median** size?

A   15.5cm

B   16cm

C   12.5cm

D   14cm

10 A builder has started to build a wall. He is using bricks with dimension of width 10cm – length 25cm – height 7.5cm.

The diagram below shows the first brick in place:

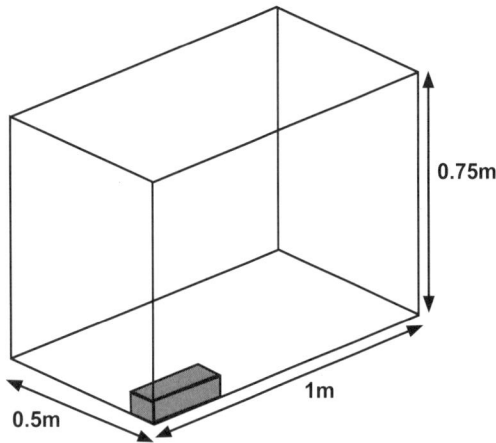

0.75m

1m

0.5m

How many bricks will it take to form this part of the wall?

A   200

B   175

C   150

D   125

11 The builder needs to make some mortar for the bricks. He uses a mortar mix ratio of 2 parts cement to 5 parts sand.

He estimates that he will need to make 3½kg of mortar. What weight of **cement** does he need to use?

A   2kg

B   1.5kg

C   1kg

D   ½kg

12 During the winter a temperature gauge reads -14°C. If the temperature **rises** 5°C, what would the temperature be?

A   -19°C

B   -9°C

C   2°C

D   9°C

13 An accountant keeps a spreadsheet on business expenditure. There was £200 balance at the start of the year:

BALANCE SHEET	Jan	Feb	Mar	Apr
Sales £s	356	305	280	324
Expenses £s	501	400	295	
Profit/Loss £s	-145	-95		-81
Balance £s	200	105	90	9

What are the **missing** expenses in April?

A   £243

B   £337

C   £391

D   £405

14 The volume of a pyramid can be calculated using the following formula:

Volume = 1/3 x Area of Base x Height

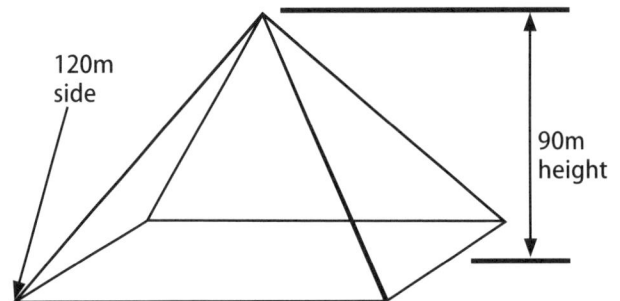

120m side

90m height

The base of the pyramid is square with all sides of equal length. What is the correct **volume** of the pyramid?

A   432,000m³

B   43,200m³

C   1,296,000m³

D   129,600m³

15 An estate agency has reduced its normal advertisement costs from £160 to £72 giving a saving of £88.

Which of the following calculations gives the saving as a **percentage** of the normal price?

A   88/160 x 100

B   160/72 x 100

C   72/88 x 100

D   88/72 x 100

16 A machinist in an engineering company can produce 240 components in eight minutes. How long will it take to produce **10,080** components?

A   10 hrs 45 mins

B   8 hrs 24 mins

C   6 hrs 8 mins

D   5 hrs 36 mins

17 A vet weighs a cat before giving it an injection. The injection is based on weight and requires **20ml** of fluid per kilogram.

The cat weighs 4lb 8oz. Using the following:

16oz = 1lb — 1lb = 0.454kg

How many **millilitres** of fluid will the cat need?

A   40ml

B   50ml

C   60ml

C   70ml

18 A woman is about to replace the water in her fish tank.

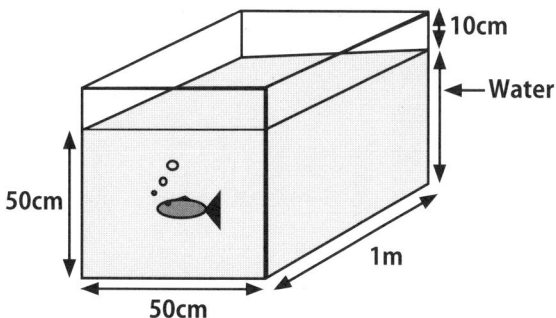

She is going to use a 10 litre bucket to refill the tank with water. Using the following:

1 millimetre = 1cm³
1,000 millimetre = 1,000cm³
1,000cm³ = 1 litre

How **many** 10 litre buckets of water will she need?

A   250

B   25

C   125

D   12.5

19 The diagram shows the distance on a digital read out:

What is the reading to the nearest 10 metres?

A   2.670km

B   2.660km

C   2.600km

D   2.700km

20 Chemicals are purchased in large 5 litre containers. The contents of the container are then poured in to smaller 300ml containers for health and safety reasons.

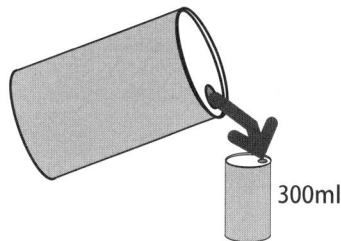

How many **full** 300ml containers will there be?

A   19

B   17.5

C   16

D   13

Chapter 3: Information and Communication Technology

At **Level 2,** a learner should be able to carry out effective searches and put together new information. You will need to present information which combines a combination of text, images and numbers in a consistent and appropriate way.

When using a spreadsheet, you will need to show you can use formulae. You will show you can develop documents by selecting and applying Information and Communication Technology skills appropriate for the task.

Learners should be able to use technology safely and care for the equipment being used. It is important to work in such as way that data is used and kept safely and not lost or damaged by viruses.

Learners will also need to observe copyright and confidentiality laws, and work healthily and safely.

It is expected that you will be able to send and receive emails.

The following Reference Sheets provide opportunities for you to review and practise the Information and Communication Technology functions needed for Key Skills.

A GUIDE TO USING COMPUTERS SAFELY

A computing system is made up of a **VDU** (**Visual Display Unit** – i.e. the screen), a keyboard, a mouse, a **CPU** (**Central Processing Unit** – i.e. the housing and its contents), a printer, a scanner, etc.

As with any other piece of equipment it is advisable to know how to use them safely.

When using computers at work you are protected by the EU law passed in **1992** called the **Display Screen Equipment Regulations**. This does not relate only to computer screens, but any other means of displaying information (data). For instance it includes such items as cash registers and calculators.

VDU (screen)

CPU (Central Processing Unit)

The Directive sets out a number of legal requirements for employers and employees and is designed to prevent some of the ill effects that some people can experience when using a computer, especially for long periods of time.

Even if you use a computer at school, college or home and are not yet working, you should be aware of how to do so safely and healthily.

So, read on and find out about the hazards so that you can avoid them. Be honest as you read through this guide, do you use your computer as safely as is recommended?

The hazards to your health

RSI (Repetitive Strain Injury) can affect many parts of the upper body causing stiffness, aches and pains. A common form of RSI, also known as Tenosynovitus, is the inflammation of the tendons in the wrist. Its symptoms can include pain, numbness and tingling in the hands and arms.

It is caused by repetitive movement such as typing and use of the mouse. It's not just typists who can get RSI, but anyone who makes the same movement over and over again for long periods of time.

RSI can be treated, but if you suffer from the symptoms it is important to act quickly to avoid permanent damage. Prevention is much easier than the cure! Make sure you sit correctly at the desk and that you take breaks regularly. The **Display Screen Equipment Regulations** state that employers must "plan work to ensure there are breaks and changes of activity – short frequent breaks are considered more useful than longer less frequent ones" (advice varies on what this should be; some sources suggest five minutes break every half hour whilst others advise 30 second micro-breaks every 10 minutes).

Place the mouse so that you do not have to reach to use it. When using the mouse keep your wrist at a level position so it is not bent. You may find that a wrist support will help to achieve the correct position.

Sight-related problems include the eyes becoming tired, dry, itchy or sore if they stare at a screen for hours at a time.

Don't spend too long looking at the monitor as you may get eyestrain. The eye's muscles are like an elastic band. They stretch to focus on the distance between your eyes and the object you are looking at. The muscles become lazy and tired if you don't keep changing the focus distance. Therefore, don't look at any distance for too long. When looking at the computer screen, regularly – (about every 10 to 15 minutes) look away to either your papers (which presumably are nearer than the screen), or to the wall in front of you (which presumably is further away than the screen).

Remember to blink. Dry eyes become itchy and sore.

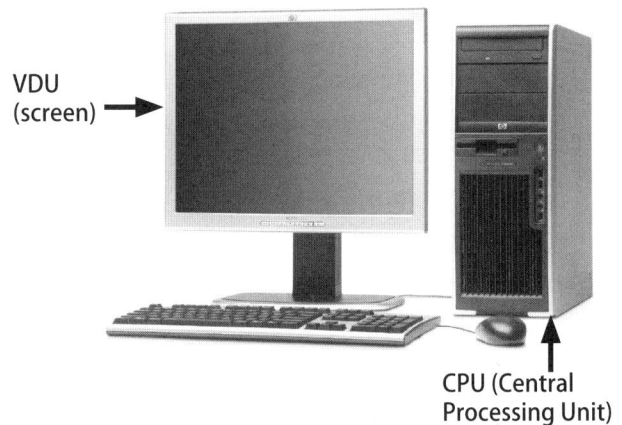

Sit the correct distance from the screen and raise the screen if necessary (*see diagram below*).

Sitting correctly at the computer

Sit up straight using a proper typist's chair that has a back support. The support should be adjusted so that it fits into the small of your back, doing what it is designed to do, give you support.

Sit about two feet (an arm's length) directly in front of your screen.

The top of the screen should be positioned so that it is no higher than your eyes, i.e. look slightly down at the screen.

If necessary, raise the screen using blocks.

Your elbows should be placed close to your body and bent at a 90-degree angle.

Place the keyboard towards the front edge of the desk, squarely in front of you, towards the edge of the desk.

Don't rest your arms or wrists on the desk.

Invest in a wrist rest so that your wrists are supported and cushioned when you are not typing.

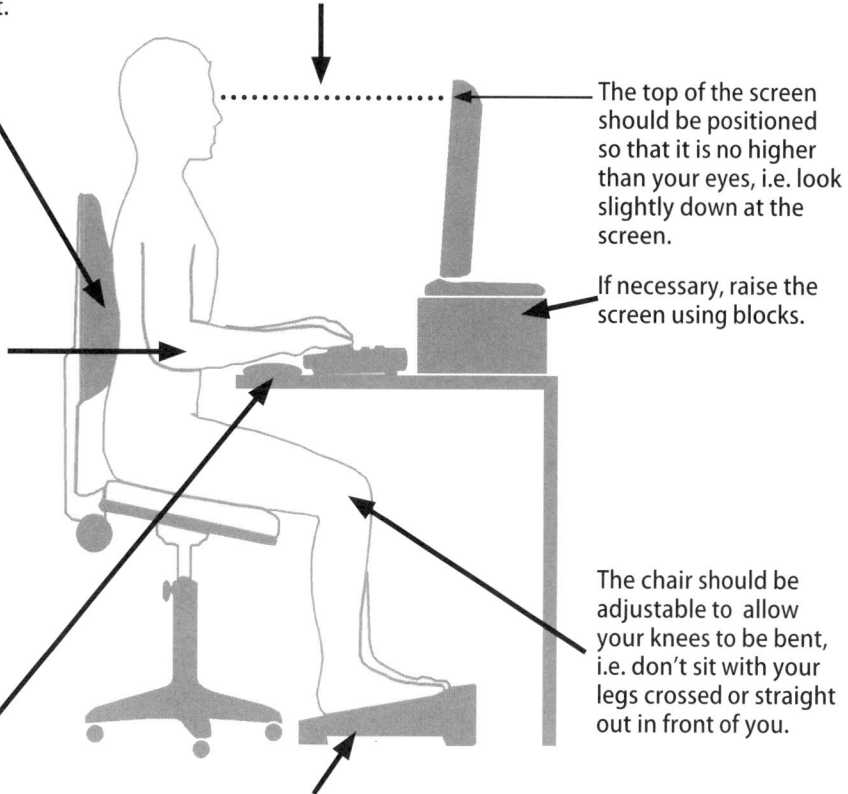

The chair should be adjustable to allow your knees to be bent, i.e. don't sit with your legs crossed or straight out in front of you.

The typing chair should be at a height so that your feet are flat on the floor. Note they should not be resting on the chair's feet or wheels.

Use a foot rest if necessary.

The hazards to the equipment

Computing equipment is expensive; it is designed to carry out tasks which help you. It, like everything else, deserves respect. Respect means treating something properly. If you respect equipment it will serve you, if you misuse equipment, one day, it will fail you.

Viruses

When you connect to the Internet you are putting your computer, and your hard work at risk from virus and hacker attacks. You need to ensure that your equipment is protected from such attacks by investing in **anti-virus protection** and installing **firewall** software as a minimum precaution.

Make sure that you regularly check for updates for these pieces of software, particularly the virus software as new viruses are created frequently.

Some on-line tips

NEVER open documents or run programs attached to emails from addresses you do not recognise. You are advised not to even open the email if possible.

NEVER give anyone you don't know, any information which will identify you. For instance, your name, address, bank details, etc.

ONLY VISIT trusted web sites.

ALWAYS use, and regularly update, your anti-virus protection package.

ALWAYS use a firewall device or software.

Some computer health tips

Always close your computer down using the correct procedure. **Don't just switch it off when you have finished typing**.

The correct procedure is:

- ✓ close the document(s) you are working on;
- ✓ probably disconnect from the Internet if you are a home user;
- ✓ take out any floppy disk you may have inserted earlier;
- ✓ click the Windows Start menu and Turn Off Computer;

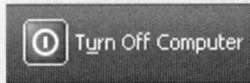

⏻ Turn Off Computer

- ✓ switch off at the mains supply.

Work station assessment

At work, if you are experiencing any problems through using a computer for prolonged periods you should ask to have a risk assessment done on the workstation. A slight adjustment could make all the difference.

REMEMBER

CARE FOR YOURSELF AND THE EQUIPMENT AND YOU WILL BOTH STAY HEALTHY

ORGANISING YOUR DOCUMENTS

If you need to keep your document for future work or for sending to somebody on a disk or by email you will need to save it.

It is good practice to organise your work into folders (also known as directories) so that you can find them easily later on. If you have a number of tasks that you are working on you can keep the files stored in appropriate folders.

For your Key Skills portfolio you are required to take screen shots of the files saved to a folder to prove that you have saved your work. It is, therefore, a good idea to keep all related files for each task in a separate folder so that anyone inspecting your work can easily and quickly see the relevant files (*see also page 145 Taking a screen shot*).

Here is an example of how you might wish to arrange your files:

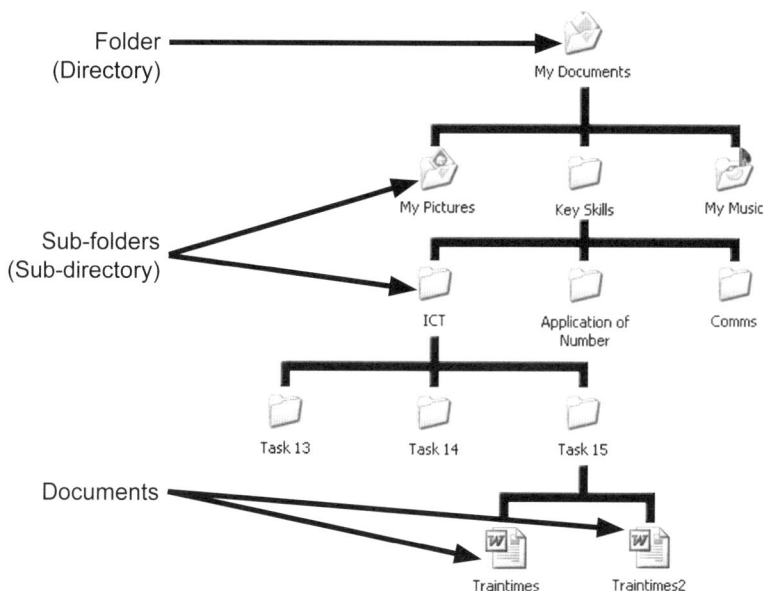

Creating a new folder (directory)

There are a number of ways in which you can create a new folder. Here you will see how to do this from within an application. The procedure is the same for all Microsoft Office applications.

➤ Select the **File** menu and then the **Save As...** menu item.

➤ If necessary change to a suitable location that you want to save the file. For example, another folder or a different disk drive.

➤ Click on the **Create New Folder** button.

➤ In the **New Folder** dialog box that is then displayed type in the name you want to call the folder.

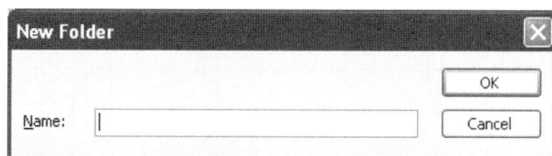

➤ Click **OK**. This will create the new folder and automatically open it ready for you to save your document (*see also page 156* for more details on saving documents).

Finding files

Sometimes, no matter how sensible you are about naming files appropriately, and placing them in folders, you can misplace a document. If this happens, you can search for different files, and file types in the following way:

▶ Click on the **Start** button on the Windows **Taskbar** and select **Search**.

The **Search Companion** wizard is displayed. There are two steps to follow in which a few questions are asked about the file you are searching for that are designed to help refine your search.

At the first step select the type of file(s) you are looking for. For instance is it a **Document** file, such as a word processing file, spreadsheet or a database?

At the second step you are asked when the file was last modified and to enter all, or part of, the document's name if you know it. Not providing a name or part of a name will, possibly, return a large number of files.

▶ Enter all or part of the name – as much as you can remember.

▶ Click on **Search**.

A list of files that have been found that match the search criteria you selected are displayed in the **Search Results** window. In this example we searched for files with "Ladybird" in the title. It has found a range of files.

Step 1

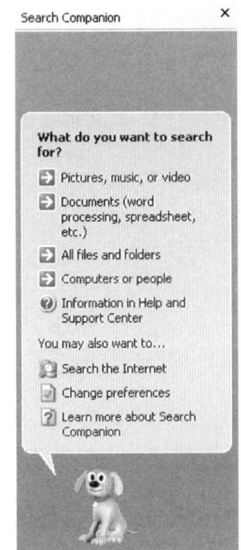

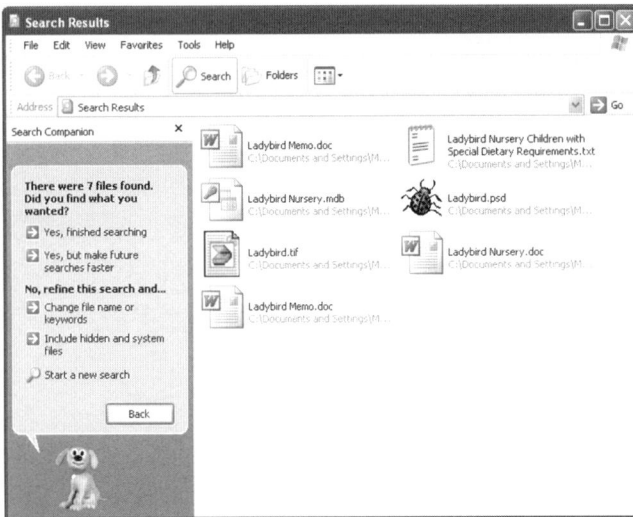

From here you can select the file you wanted and either open it, print it, copy it or move it.

If, however, you have not found what you have been looking for, click the **Back** button to change your search criteria.

Close the **Search Results** window by clicking on the **Close** button on the window.

Step 2

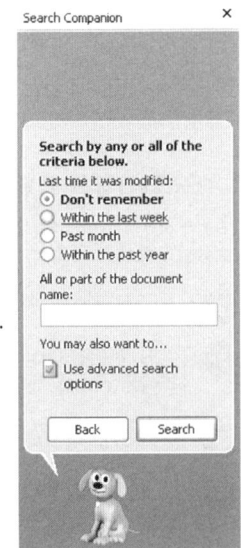

Remember

Name and save files appropriately and logically!

Fun tip

Right click on the dog to make him do tricks or change the animation.

Using wildcards when searching for documents

Using the **Search Companion** method described on *page 142* should, under normal circumstances, find the files you are looking for without having to use Wildcards. However, this technique is useful to know and you are required to be familiar with these at **Level 2**.

Wildcards are characters that can be inserted in a search term and are particularly useful when:

- ✓ you are not entirely sure of the name of a file;
- ✓ you want to find different variants of that name;
- ✓ you want to find all files of a certain type.

There are two wildcards to use when searching for files or folders: question mark (**?**) and the asterisk (*****).

- ✓ An asterisk (*****) is used as a substitute for **zero or more** characters.
- ✓ A question mark (**?**) is used as a substitute a single character in a filename.

Here are some useful ways in which you could use these wildcards:

Wildcard	Searches for	Possible search results
ladybird?.doc	Any filename starting with ladybird and ending with just one character	ladybird1.doc, ladybird2.doc, ladybird3.doc, ladybirda.doc, ladybirdb.doc, ladybirds.doc
ladybird??.doc	Any filename starting with ladybird and ending with just two characters	ladybird10.doc, ladybird2a.doc, ladybirds1.doc, ladybird-2.doc, ladybird_1.doc,
lady*	The word lady and any amount of text after it	lady.doc, ladybird.doc, ladybirds.doc, lady.jpg, lady bird.xls, lady jane.rtf, etc.
mark*file.doc	Any filename starting with **mark** and ending in **file.doc**	marks homework file.doc, mark's personal file.doc, mark'sfile.doc, etc.
ladybird.*	Any file type called **ladybird**	ladybird.doc, ladybird.jpg, ladybird.tif, ladybird.mdb, ladybird.txt, etc.
.	Any file of any type	All the above and more!

USING HELP SYSTEMS

You will often find times when you need to find a solution to a problem you have when using software, or you need a quick reminder of how to carry out a procedure. All good software applications have a help system supplied with the software and a large number of these work in a similar way. Microsoft software products all work in a similar way depending upon the version you are using.

Microsoft Office 2003 has a number of ways of finding information to help you use the software including files stored on your computer and, if you are connected to the Internet, the latest help files are searched for on the Microsoft website via your Office application.

In the following example we will show a basic method of how to find help on inserting a footer using Excel.

> Launch Excel.

> To launch help do **one** of the following:

Select the **Help** menu and then the **Microsoft Excel Help** menu item.

Click **Help** button from the **Standard toolbar**.

Press the **F1** key (this is common to most software programs).

Excel Help is displayed in the **Task pane**.

> Type **insert footer** into the **Search for** box and click on the **Start searching** button.

If you are connected to the Internet the following will be displayed in the **Task pane**:

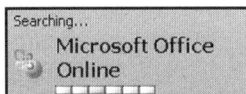

Searching...
Microsoft Office
Online

The initial Excel Help Task pane

The search results from the Microsoft website or from the files on your computer are displayed by relevance in the task pane (*see Figure right*).

> Click on the heading that you feel is the closest match to your query. In this example we have chosen the first result: **Add headers and footers for printing**.

The help file is then opened in a new window:

Show the help text by clicking on the link.

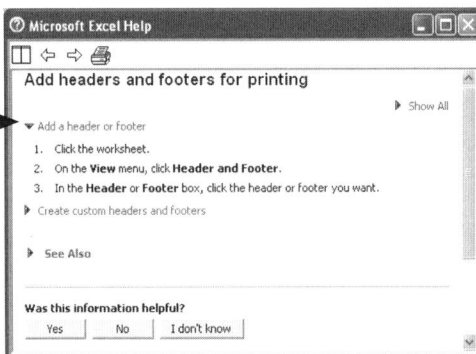

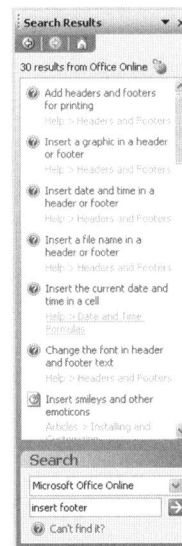

Microsoft Excel Help

Add headers and footers for printing

▶ Show All

▼ Add a header or footer
1. Click the worksheet.
2. On the **View** menu, click **Header and Footer**.
3. In the **Header** or **Footer** box, click the header or footer you want.

▶ Create custom headers and footers

▶ See Also

Was this information helpful?
Yes   No   I don't know

The search results

> If you are presented with further choices within the help file (as above) click on the link to expand the view of the text.

TAKING A SCREEN SHOT (SCREEN PRINT)

For Key Skills in ICT you will need to provide evidence for the assessors that you have successfully carried out the required tasks. To do this you will need to produce a copy of all the folders/files you created in a task. This may be in the form of a **screen shot** of the filenames along with your name, the date and the title of the task.

You may find screen shots useful for other purposes; many of the pictures in this book, for example, were made using screen shots.

It is a very easy to do and can be used with a number of applications. As you need to add information about the tasks we suggest that you use **Microsoft Word**:

- Open the folder containing the files you need to print a copy of.

- Maximise the folder to fit the screen by clicking on the **Maximize** button (this will save you having to crop the image and if your folder contains a large number of files it will show a greater number of them).

- Press the **Print Screen** key on your keyboard. This takes a snap shot of the screen and places it into the Windows' **clipboard**.

- **Open** a new document in Word and position the cursor in the document.

- Click the **Paste** button from the **Standard Toolbar**.

The image will now be inserted into the new document.

- **Resize** the image as necessary to fit on the page.

- Add **your name**, **the date the document was created** and the **title of the task**.

- **Print** a copy of the document.

Tip

You may want to change the way the files are displayed in the folder to show more files or more information about the files.

- Select the **View** menu of the folder you want to include in a screen shot.

- Chose one of the following options:

  Thumbnails
  Titles
  List
  Details

Experiment to see which gives you the best view.

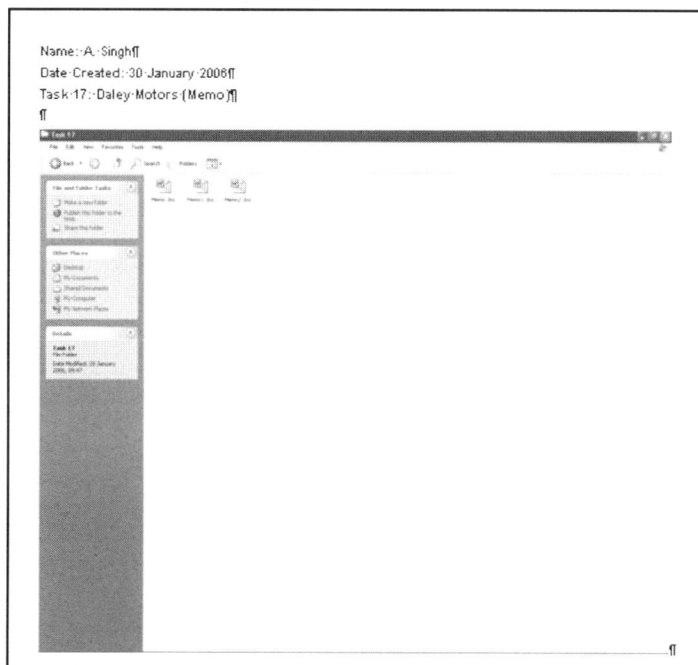

How your final document might look.

145

USING THE INTERNET AND EXPLORING THE WORLD WIDE WEB

What is the Internet?

The word **Internet** is abbreviated from **internetwork**, which literally means a network of interconnected networks. It allows computers to communicate with each other and exchange information. By using your computer to access the Internet you are able to view information from the huge global "library" of electronic information known as the **World Wide Web** (**www**) and communicate with others, for example by sending and receiving **email**.

To view web pages and "surf" the Internet you need to use a web browser, for example **Internet Explorer**, **Netscape** or **FireFox**. In our examples we use **Internet Explorer**.

Using the World Wide Web

This huge global library is made up of millions of websites that are, in turn, made up of web pages. A website might contain just one page or millions of pages. To move around a website you use **hyperlinks**. Hyperlinks are used on text (usually identified in a different colour to the text) and on pictures (graphics).

The easiest way to identify a hyperlink is to look at your mouse pointer. As you place it over a hyperlink it changes to a hand like this:

Most websites have a **Uniform Resource Locator** (**URL**), which is often referred to as a **web address**. URLs are used in hyperlinks to link pages and other websites together and can be used in the browser address bar to access a website directly. It is very important, when using a URL to access a website, to type it in correctly otherwise you will not find the correct website. Think of it like a telephone number if you dial the wrong number you either speak to the wrong person or don't get through at all!

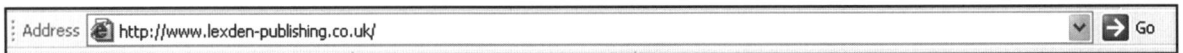

Address	http://www.lexden-publishing.co.uk/	✓	→ Go

Accessing a web page using its web address

To access the Internet using Microsoft's **Internet Explorer**:

⊙ Double-click the icon which can be found on the desktop or select it from the **Start** menu and then **All Programs**.

The first page you will see is known as the **Homepage**. This is set in the browser and can be changed to a more useful website if necessary. Typically the homepage might be the school or college's website or intranet site, or your favourite search engine site such as **Google**.

To go to another website:

⊙ Click in the address box. This should highlight the current URL and you can type the new URL directly into the address bar. If it doesn't highlight the current URL then use the **delete** or **backspace** keys to delete it.

⊙ Type in **www.lexden-publishing.co.uk/keyskills**.

⊙ Click the **Go** button (or press the **Enter** key) to visit our website.

⊙ Move your mouse around the page to find the hyperlinks.

> **Tip**
>
> You don't need to type in **http://** as the browser will do this for you.

Parts of URLs mean different things and can give you clues about the organisation.

This is the **Domain name** (a company name or the name of an **Internet Service Provider (ISP)**).

The company is based in the UK.

www.lexden-publishing.co.uk

www lets the browser know that the page is a World Wide Web page.

It is a company.

Other organisational type suffixes you might see include:

.ac   An academic body, school, college of university

.co   A business

.com   A commercial organisation (intended for American companies, but is used globally)

.edu   An American academic body or institution

.gov   A government agency

.net   Internet Service Provider (intended initially for ISPs, but other companies/organisations now use this as well)

.org   An organisation

.sch   A school

Country suffixes you might see include:

.de	Germany	**.pt**	Portugal
.jp	Japan	**.nz**	New Zealand
.uk	United Kingdom		

Useful web browser buttons

Back button: When you have visited more than one web page you can use the **Back** button to return to a previous page. Keep clicking this to take you through each page.

Forward button: Once you have used the **Back** button you can then use the **Forward** button to go the other way.

Remember though that this is limited in that you can only go back or forward a certain number of times held in the memory.

Refresh button: Some web pages get updated regularly, such as a news page. Click on the **Refresh** button to reload the page to see the latest changes. You can also use the **Refresh** button if a page is not loading up correctly or is very slow as it sometimes fixes the problem.

Home button: If you wish to return to the Home page set in your browser click the **Home** button at any time.

Information and Communication Technology: Reference Sheets

3

Finding information – using the world wide library of information

The web is huge with millions upon millions of web pages and without the aid of search engines it would be very difficult to find the information you wanted. Fortunately there are a number of search engine websites that spend the whole time trawling the Internet looking for new information to index that you can then use to help find the information you want.

One of the most popular search sites is **Google**.

⊙ Open **Internet Explorer**.

⊙ Type in **www.google.co.uk** into the **Address bar**.

⊙ In the search box type in what you are looking for. In the example above we are looking for information on **Madeira Island**.

⊙ Click **Google Search**.

The search results will then be displayed.

You will get a choice of lots of sites to visit – like lots of books on the shelves of the library on a certain topic.

Tip

The search engine software places the results in order of what it believes to be the most relevant to the words you searched for.

⊙ Read the brief description of the search result to help you select the site that best describes what you are looking for.

⊙ Click on the **hyperlink**.

That website is then displayed in the browser. If it is not what you are looking for, click on the **Back** button to return to the search results at Google and select another document to view.

Tip

Press the **Shift** key when you click on a hyperlink to open the website in a new window. This will leave the search results available for you to view whilst you look around another site.

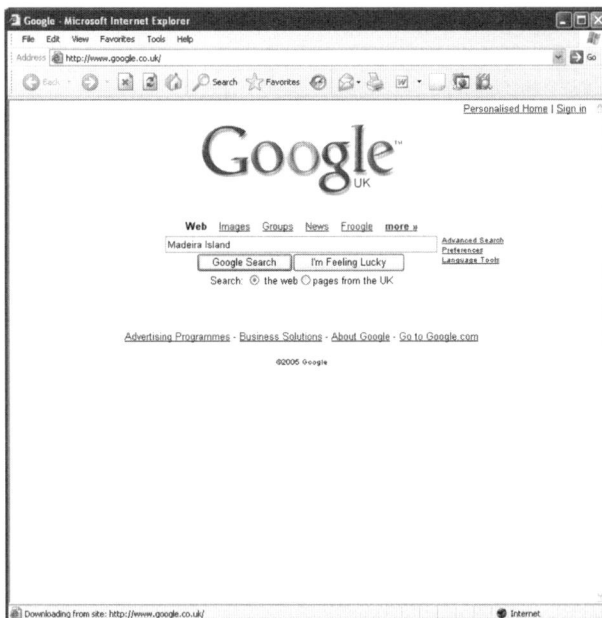

Having difficulty in finding what you want? Refine your search

As you will soon discover, search engines will find thousands upon thousands of documents with the words you searched for, many of which are irrelevant.

The secret to getting relevant information is to search correctly.

For instance, if you are looking for details of **wind farms and climate change**, you will retrieve results for: wind farms and climate change, wind farms, wind, farms and climate amongst your results.

To refine your results try putting the search phrase into quotation marks. This will find documents with the exact phrase **"wind farms and climate change"**.

Many search engines have different methods of refining searching and many, such as Google, have an advanced search page. Look for this "advanced search option" if you can't find what you are looking for in the normal way.

Printing text and images

To print the entire web page you are interested in:

⊙ Click the **Print** button.

To print a part of the web page you are interested in:

⊙ Highlight a section of the text (and images if you wish).

⊙ Select the **File** menu and then the **Print...** menu item.

⊙ In the **Print** window select the **Selection** option under **Print Range**.

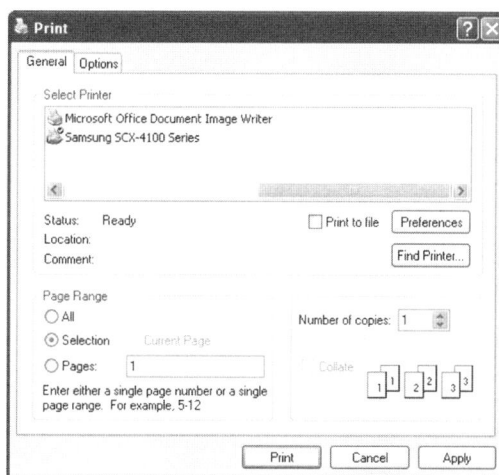

⊙ Click **Print**.

To print an image only:

⊙ Right-click the image and select **Print Picture** from the pop-up menu.

⊙ Click **Print** in the **Print** window.

Copying text and images

You can copy text and images into other applications such as **Microsoft Word** and **Excel**.

- Select the text and images to be copied from the web page.
- Select the **Edit** menu and then the **Copy** menu item.
- **Open** a document in **Microsoft Word** (for example).
- Position the cursor where you want to insert the text.
- Select the **Edit** menu and then the **Paste** menu item.

Saving images

You can save images as files onto your computer.

- Right-click the image and select **Save Picture As...** from the pop-up menu.
- Choose a folder to save the image into.
- Choose a name of the image if you are not happy with the one used by the website.
- Click **Save**.

Words of caution

Do not assume that everything you access on the World Wide Web is true. Anyone can create a website and the content is not necessarily accurate or unbiased. If you set up a website, remember that legally you are responsible for its content.

Whilst the content of the World Wide Web is available for everyone to access, remember material is protected by **copyright** laws. You can access the information but you must not claim that information as your own. Just like a library book, you can not take the work of another person and claim it as your work.

Be careful when accessing the Internet and downloading files and information – this is how computer viruses are spread. Only visit trusted sites and make sure your computer has a virus protection package installed before you access the World Wide Web or use email packages.

PRACTICE EXERCISE 1 – FORTH RAILWAY BRIDGE

In this task you are going to find some information about the history of the Forth Railway Bridge in Edinburgh.

1   Open your web browser and go to your favourite search engine website.

2   In the search box enter the words **Forth Railway Bridge**.

3   Begin the search process.

4   Access one of the pages from the search results that might give you the information you want and print the result.

5   Refine the search to **Forth Railway Bridge History** and begin the search.

6   Note down the differences in the documents available, saying whether this search seems to be more relevant.

7   Access one of the pages from the search results.

8   **Save** an image from the web page (remember where you save it to).

9   Select a small section of text that will go with the image.

10   Open a **new** document in **Microsoft Word**.

11   **Paste** in the text you copied from the website and **insert** the image you saved previously.

12   Experiment with the layout until you are happy with it and then **print** a copy.

13   **Save** the document with an appropriate filename.

14   **Close** the Word document.

EMAIL

Email is an "electronic postal system". Messages can be sent to and from anyone with an email address extremely quickly, often in minutes or even seconds!

An email can have files **attached** to the message. This is useful if you want to send a copy of a document you have written, such as a report, or an itinerary, or if you want to send photographs to someone.

Email addresses

An email address is made up of two main parts joined by a @ sign and takes the form:

user_id@domain_name

an example of a typical email address might be:

pollypage@nostampmail.co.uk

User ID   'at'   Domain name

> **Tip**
>
> When you read an email address to someone say it as, in this instance:
>
> **pollypage at no stamp mail dot co dot uk**

Domain name

This is a unique name that can be used to identify a computer connected to the Internet. They are often similar to a company/organisation's name or even a product name. They are used in URLs and in email addresses. Personal email addresses quite often use an **Internet Service Provider's (ISP)** domain name (e.g. ntlworld.com) or an email service (e.g. yahoo.com) as part of the address.

User ID

A user ID must be unique within a domain. For instance if there are two people named Polly Page using the nostampmail.co.uk domain and one already has the email address pollypage@nostampmail.co.uk the other would have to use a unique user ID such as polly.page@nostampmail.co.uk or pollypage1966@nostampmail.co.uk.

Setting up a web-based email account

There are many different email account providers. Your tutor will advise you which ones are available in your college or school. If there is no special system, then a free email account may be set up using **Yahoo**. Once you have set up an account you can access email through the Internet, anywhere, anytime.

To set up an account:

▶ Open **Internet Explorer**.

▶ Type in **uk.yahoo.com** into the **Address bar**. (Note that www is not used in this instance.)

The web page will look similar to this one:

Yahoo's sign in form

▶ Choose **Mail** button to the right of the Yahoo banner.

When the next web page loads you will see the **Sign in** form (*see above right*). Here you can create a new account.

▶ Click on **Sign Up** and follow the on screen instructions.

You will need to think of a Yahoo ID (a user name that will be unique to you and will be the **user ID** part of your email address) and choose a password.

If you have a common or popular name, you may have to be a little creative otherwise someone else may have been allocated it already.

Once you have completed the sign up process you will be ready to send your first email.

Test your new email address

Once you have set up your account you will want to test that it is working correctly.

Sign into your email account using the screen shown on *page 152* and compose a new message to another member of your group. You can decide upon the content of the email as long as the text is sensible and polite.

> **Note**
>
> You will need to know the email address of the person you are sending to.

When you have finished writing the email you will need to select **Send** for the message to go to its recipient.

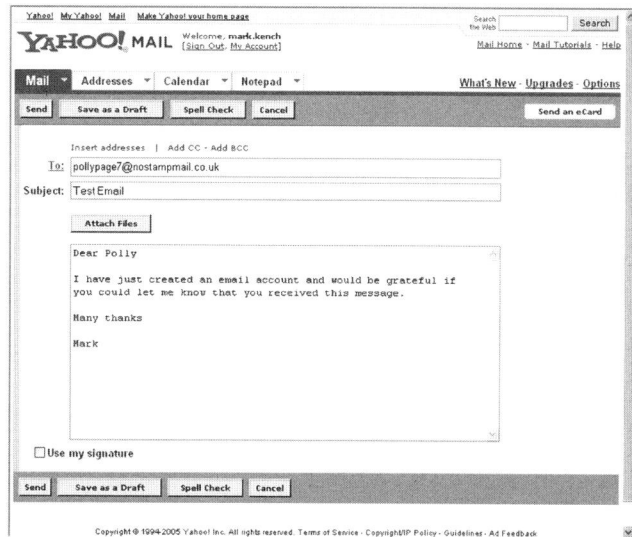

PRACTICE EXERCISE 2 – EMAIL DOS AND DON'TS

In this task you are going to search the Internet for guidance on the etiquette surrounding sending emails.

Remember: emails are not just communications between friends (informal, chatty documents), they can be sent from home to a company (perhaps applying for a job, or making a formal complaint about goods or services) and, of course, they are used in companies to send messages to other employees and employers.

Once you have sent an email it is almost impossible to retrieve it.

For this reason it is important to be polite and accurate.

Because your Key Skills course requires you to send and receive emails, your task now is to search the Internet for **Email dos and don'ts** and print out a copy for your file.

1. Open your web browser and go to your favourite search engine website.

2. In the search box enter words which should help you locate the information you want, for instance **Email dos and don'ts**.

3. Begin the search process.

4. **Print** a copy of the first results page.

5. Access one of the web pages from the search results that may give you information you need.

6. If it proves to be just the information you need **print** a copy of it for your file.

7. If it is not quite what you are looking for, begin the search again, refining your search text until you do find something relevant and can print it.

Sending and receiving emails with an attachment

One of the great things about email is the ability to be able to send files, such as documents or images, to colleagues, friends and family. The process of sending attachments is fairly similar in most email applications. In this example we are using **Microsoft Outlook Express**.

Sending attachments

▶ Launch **Outlook Express**.

⌧ **Outlook Express**

▶ Create a new email message by clicking on the **Create Mail** button.

Create Mail

The **New Message** window is displayed ready for you to compose your email.

▶ Click on the **Attach** button.

Attach

The **Insert Attachment** window is then displayed.

Tip

Use the **Ctrl** or **Shift** key to select more files at the same time.

Insert Attachment

Look in: Task 2

Ladybird 1.doc
Ladybird 2.doc

Click here to browse to the file you want to attach to your email.

Tip

It is very easy to forget to add an attachment. Add the attachment first then write your message.

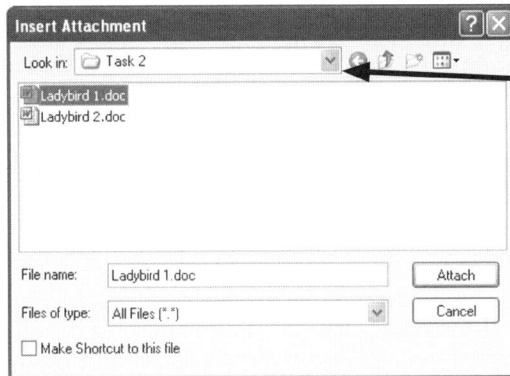

File name: Ladybird 1.doc   Attach

Files of type: All Files (*.*)   Cancel

☐ Make Shortcut to this file

▶ Browse to the file you need to attach to your document by clicking on the down arrow on the **Look In** drop-down list.

▶ When you have located the file click on it so that the filename appears in the **File name** text box.

▶ Click on the **Attach** button.

The **Insert Attachment** window is closed and you will now see the filename of your selected document displayed in the **Attach** text box.

▶ Click on the **Send** button when you are ready to send the message.

Send

The attachment is shown here. If it is the wrong file select it and press the **Delete** key.

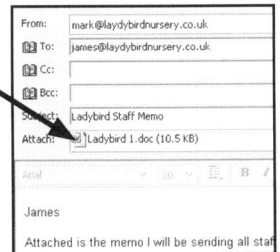

From: mark@laydybirdnursery.co.uk
To: james@laydybirdnursery.co.uk
Cc:
Bcc:
Subject: Ladybird Staff Memo
Attach: Ladybird 1.doc (10.5 KB)

Arial   10

James

Attached is the memo I will be sending all staff

Receiving attachments

When you receive an email with an attachment either:

▶ Open it by double-clicking on the attachment icon or name in the **Attach** text box (*see Figure above*).

▶ A warning dialogue will be displayed (*see right*). If you are happy that the file will not cause your system any problems, select **Open**.

Or

▶ Save the attachment by selecting the **File** menu and then the **Save Attachments** menu item and decide where you want to save the attachment. (*See also page 156 Saving your Work.*)

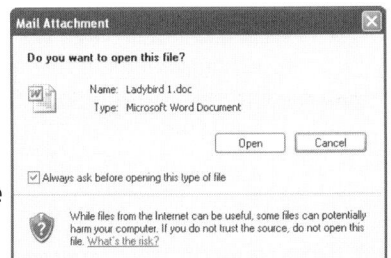

Mail Attachment

Do you want to open this file?

Name: Ladybird 1.doc
Type: Microsoft Word Document

Open   Cancel

☑ Always ask before opening this type of file

While files from the Internet can be useful, some files can potentially harm your computer. If you do not trust the source, do not open this file. What's the risk?

WORD PROCESSING

What is a word processor?

Word processors are probably the most widely used piece of software in the world. Word processors are used in business and the home because they have powerful editing features such as spelling and grammar checking; can be used to format text in different styles; and can combine images in the documents for writing letters, reports, newsletters and even books.

Microsoft Word is the word processing application used in this book.

Creating a new document in Word

When you first open **Microsoft Word**, a blank document is normally created by default. If it is not click on the **New Blank Document** button.

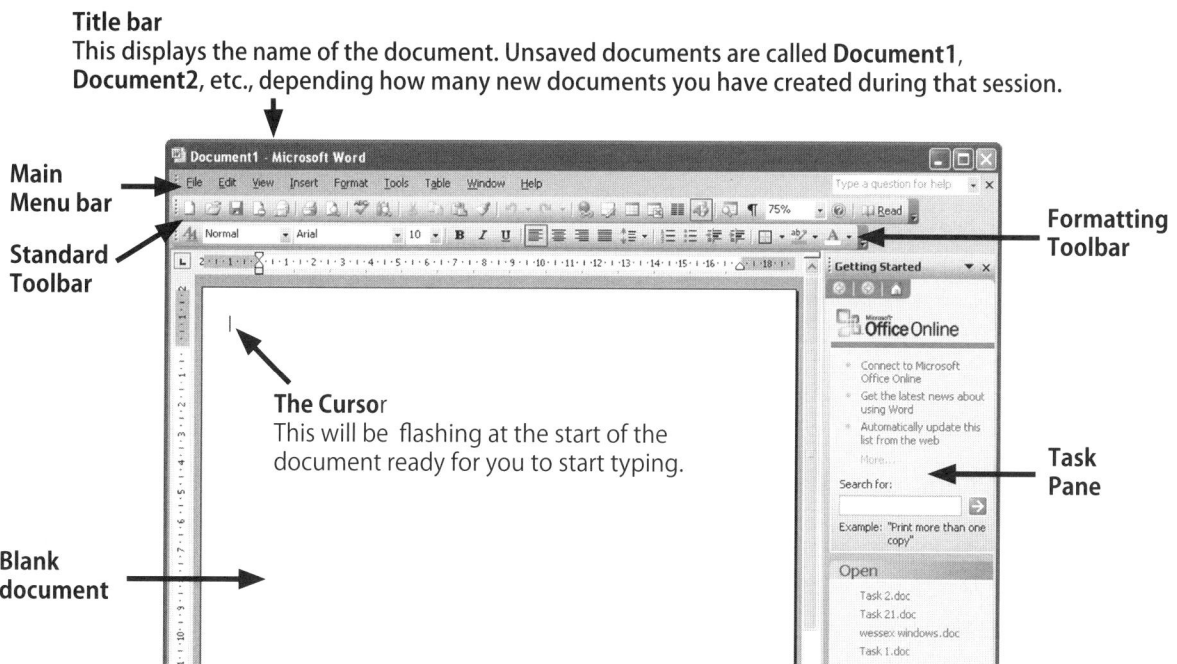

Title bar
This displays the name of the document. Unsaved documents are called **Document1**, **Document2**, etc., depending how many new documents you have created during that session.

Main Menu bar

Standard Toolbar

Formatting Toolbar

The Cursor
This will be flashing at the start of the document ready for you to start typing.

Task Pane

Blank document

Entering text

When you have created a new blank document the cursor should already be flashing at the beginning of the document ready for you to start typing. If it is not, click on the document. When you type, Word will automatically wrap the text to begin a new line. However, if you need to begin a new line (for example, to start a new paragraph), press the **Enter** key on the keyboard.

- When you work through the exercises in this book you do not have to have the lines end in exactly the same place as the examples.
- Generally, two spaces are entered after a full stop and one after other punctuation marks.
- Check your work thoroughly and correct errors.
- Use the **Spelling and Grammar** checker and read through your work before moving on to the next step. Make sure that there are no errors.

Saving your work

Save

If you intend to keep your work then you need to **save** it. Do this often as you work to reduce the risk of losing your work. When saving your work for the first time:

⊳ Select the **File** menu.

⊳ Select the select the **Save** menu item to display the **Save As** window.

⊳ Give your document an appropriate filename and choose a folder or drive to save your file to, e.g. a floppy disk or network drive.

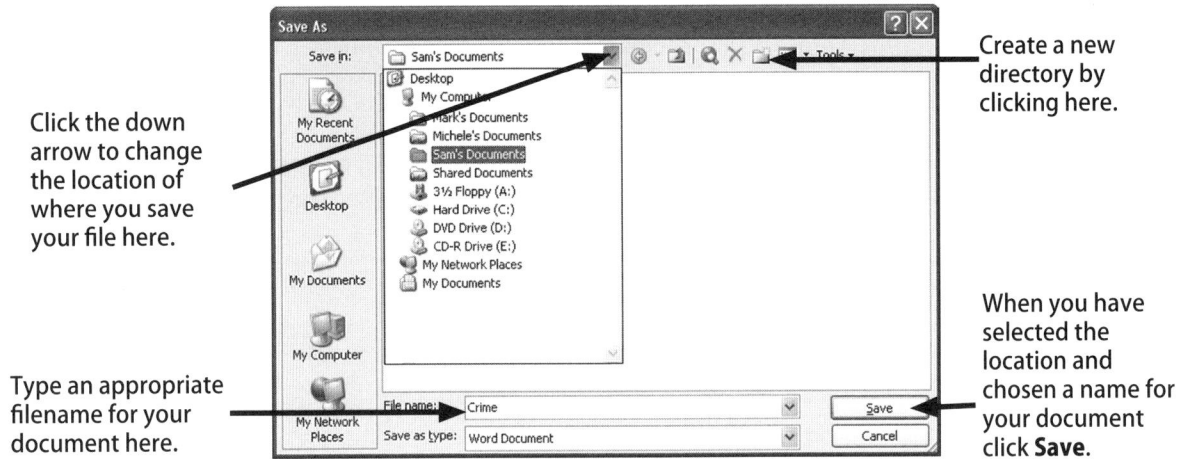

Click the down arrow to change the location of where you save your file here.

Create a new directory by clicking here.

Type an appropriate filename for your document here.

When you have selected the location and chosen a name for your document click **Save**.

When you save your work again simply select the **File** menu and **Save** again or click on the **Save** button on the **Standard Toobar**.

The **Save As** window is not displayed in subsequent saves.

Save As

You will often need to save different versions of your documents with new filenames. To do this:

⊳ Select the **File** menu.

⊳ Select the **Save As** menu item to display the **Save As** window.

⊳ Give your document an appropriate filename and choose a folder or drive to save your file to, e.g. a floppy disk or network drive.

Fonts

Fonts are a complete collection of letters and symbols belonging to a **typeface** that can be displayed in any **point size** or **weight** (**bold**, **italic**, etc.). A **typeface** is the style and design of a **font**.

Font Size describes the size of the text and is also known as the **Point Size**.

Choosing the right typeface for your task is important and you need to think about who your document is going to be read by. If you are writing a letter for a job, it would not be advisable to use **Comic Sans** or a **Script**. You might, however, chose a more formal typeface such as **Times New Roman** or possibly **Arial**. Try not to mix too many typefaces in your document.

This typeface is in 20 point Arial bold

This typeface is in 18 point Times New Roman

This typeface is in 16 point Comic Sans

This typeface is in 12 point Script

To change the **Font** and its **attributes** (**size**, **weight**, etc.) highlight the text to be changed and select one or more of the options available from the **Formatting** toolbar (*see below*).

Font (typeface)

Click on the down arrow to select a font from the drop-down list. Scroll down the list if necessary.

Font attributes

Click on one or more of these buttons to apply the required style. Clicking the button again will remove the formatting.

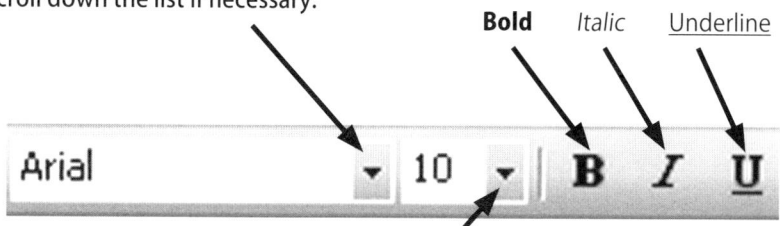

Bold   *Italic*   <u>Underline</u>

| Arial | ▼ | 10 | ▼ | **B** | *I* | <u>U</u> |

Font Size (Point Size)

Click on the down arrow to select a font size from the drop-down list. Scroll down the list if necessary. You can also type a size in if you want to if it's not shown in the list.

Line spacing

There are three main types of line spacing:

Single

> Hickory Dickory Dock, the mouse ran up the clock, the clock struck one the mouse ran down. Hickory Dickory Dock.

Double

> Hickory Dickory Dock, the mouse ran up the clock, the clock struck
>
> one the mouse ran down. Hickory Dickory Dock.

One and a half

> Hickory Dickory Dock, the mouse ran up the clock, the clock struck
>
> one the mouse ran down. Hickory Dickory Dock.

If you are asked to change line spacing, choose the **Format** menu and then the **Paragraph** menu item. Then alter the **Line spacing** settings in the **Paragraph** window that is displayed (*see page 159*). Alternatively, you can use keyboard shortcut. For example, place the cursor in the text to be changed (or highlight the text if there is more than one paragraph to change), then use **CTRL+2** for **double line** spacing and **CTRL+5** for **one and a half line** spacing. The document will usually be in **single line spacing**, but if not and single is required choose **CTRL+1**.

When entering text, there should **always** be one clear line space between paragraphs whatever type of line spacing is used. You will notice this tends not to be the case in newspaper, or magazine, articles. However, leaving a line space between paragraphs is accepted practice for business documents and it is a good idea to present your work in this way.

Find and Replace

To use **Find and Replace**:

⊙ Select the **Edit** menu.

⊙ Select the **Replace** menu item to display the **Find and Replace** window.

⊙ In the **Find what** box, type the word that is already in the text that you want to change.

⊙ Enter the word to replace it in the **Replace with** box.

⊙ Click on **Replace All**.

⊙ **Word** will automatically search your document and tell you how many it has found and replaced. Click **OK** when completed.

Deleting text

To delete a sentence or paragraph, first highlight the text with the mouse and then press the **Delete** key. Check to ensure that all relevant full stops are included in your highlighted text.

Inserting text

Find the location in the text where you are asked to insert some text. Click the mouse so that the cursor is flashing at the correct location and then type in the new information. Word will automatically move the other text down as you work.

Moving text (Cut, Copy and Paste)

When asked to move text from one place to another use the **Cut** and **Paste** buttons on the **Standard Toolbar**.

⊙ Highlight the text to be moved.

⊙ Use the **Cut** button to cut from its existing position.

⊙ Click with the mouse to insert the cursor in the new position.

⊙ Click the **Paste** button to place the text.

> **Note**
>
> To copy text click on the **Copy** button on the **Standard Toolbar** rather than the **Cut** button.

Text alignment

There are four types of alignment:

Align Left – this aligns text so that it is straight on the left-hand side and ragged on the right.

Center – this centres text across the column. Both sides of the text are ragged.

Align Right – this aligns text so that it is straight on the right-hand side and ragged on the left.

Justify – this is when all of the text is completely straight on both sides of the column.

Note

The text you want to change must have the cursor placed in it (or be highlighted if there is more than one paragraph to change) before making one of these selections.

Indenting text

Paragraphs can be indented to make them stand out from other paragraphs. To do this:

▶ Place the cursor in the paragraph to be changed (or highlight the text if there is more than one to be changed).

▶ Select the **Format** menu and then the **Paragraph...** menu item.

▶ In the **Paragraph** window under **Indentation**, change the **Left** or **Right** setting to match your requirements.

Change indentation here

Change line spacing here

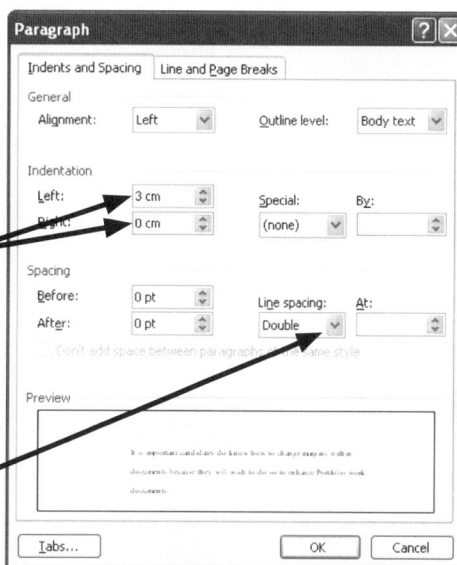

Numbered and bullet lists

It is often useful to create numbered or bullet lists. For example:

1. Chocolate
2. Sweets
3. Burgers

- Chocolate
- Sweets
- Burgers

This is done by placing the cursor in the text you want to change and then selecting either:

for a numbered list; or

for a bullet list from the **Standard Toolbar**.

Changing margin settings

Word will already have pre-set margins (called **default**). To alter these settings:

▶ Select the **File** menu and then the **Page Setup...** menu item.

▶ In the **Page Setup** window change the margin settings by typing the new measurement or by using the up and down arrows.

▶ When you are happy with the new measurements click **OK**.

Type the new margin settings...

...or click the up or down arrows to change the settings.

Default margin settings

The Default margin settings are 2.54cm for **Top** and **Bottom** and 3.17cm **Right** and **Left**.

Changing the settings and clicking the Default button will set the Default settings to the new selection for all documents.

Printing

Ensure that you print your document when asked to do so. By the end of a task, you may have more than one copy of the text. It is good practice to enter your name at the bottom of the page so that there is no confusion at the printer as to whom the work belongs. To print a document:

▶ Select the **File** menu and then the **Print...** menu item. The **Print** window is displayed.

Prints all of the pages in the document.

Prints just the page of the document shown in the window.

Prints just pages you want to print. For example, 1,2,5 would print just those three pages, whereas 1-5 would print all five pages.

Change the number of copies needed here.

▶ Click **OK** when you have made your selections.

PRACTICE EXERCISE 3 – CREATING A DOCUMENT

1   Open Word and create a **New** document.

2   Set your margins as follows:

       Top and **Bottom**   3cm

       Left and **Right**   2cm

3   Set the line spacing to **single** and **justify** the text.

4   Key in the text shown below underlining, centering and emboldening where indicated.

5   Insert your name at the end of the text to identify it.

6   **Print** a copy of the document.

7   **Save** the document, using the filename **Zodiac**.

THE WATER SIGNS OF THE ZODIAC AND THEIR MEANINGS

PISCES

Pisces is the last sign in the astrological calendar and is represented by two joined fish, swimming in opposite directions. It is the wisest of the signs, driven by emotion.

Its strengths are that it is a sign of love and with its selflessness and idealism, it is also a sign of salvation.

Through its example, Pisces teaches the other signs that many of life's problems are unimportant. The real world is that of the spirit, the realm of mystics, dreamers and visionaries.

CANCER

The Sun enters Cancer on 21st June, the day when it reaches its furthest distance from the Equator, thus giving the shortest day in the Southern hemisphere and the longest day in the Northern hemisphere.

People born under this sign of the zodiac are shy and sensitive, yet confident and ambitious.

Its strengths are that it is a sign dedicated to achieving success and no challenge is too great.

SCORPIO

Scorpio is a deep, dark, mysterious sign, renowned for its intensity and passions.

Its strengths are that it is the sign of the healer and is skilled at treating emotional wounds. It is also loyal to its friends and has a willingness to help those in need.

Creating a table

A table is a made up of one or more **columns** and **rows**, and these are made up of **Cells**.

A table is useful for displaying information like that illustrated below:

Gloss Paint Colours	Matt Paint Colours	Satin Paint Colours
Almond Green	Almond Green	Leaf Green
Sunshine Yellow	Sunburst Yellow	Daffodil Yellow

You can see the text is easy to read. It is simple for the reader to look down the **Gloss Paint Colours** column and see all that is available.

REMEMBER

Rows go across: **you row across a river**.

Columns go up and down like **Nelson's Column**.

- Select the **Table** menu.
- Select the **Insert** menu item and then **Table...**
- In the **Insert Table** window type in the number of columns and rows.
- Click **OK**.

In our paint colours example there are three columns and three rows. Below is an example of what you will have on screen:

Type the information in each **Cell**, beginning with the top left and using the **Tab** key to progress from cell to cell along the row and down to the next column.

Inserting rows in a table

- Place the cursor in any cell adjacent to where you want the new row inserted.
- Select the **Table** menu.
- Select the **Insert** menu item.
- Select either the **Rows Above** or **Rows Below** menu item.

A new row is then inserted.

Tip

By highlighting a number of rows (e.g. four) and following this procedure, an equal number (e.g. four) of rows can be inserted.

Inserting columns in a table

To insert a column in a table:

▶ Place the cursor in a cell (or highlight the column if you prefer, *see figure below*) next to where you want to insert the new column. In this example we are going to insert a column to the right of the **First Name** column.

Surname¤	First·Name¤	¤
Jackson¤	Mark¤	¤
Kennedy¤	Pauline¤	¤
McIntosh¤	Stuart¤	¤

▶ Select the **Table** menu.

▶ Select **Insert** menu item.

▶ Select **Columns to the right** from the pop-out menu.

A new column is inserted to the right ready to enter text into.

Surname¤	First·Name¤	¤	¤
Jackson¤	Mark¤	¤	¤
Kennedy¤	Pauline¤	¤	¤
McIntosh¤	Stuart¤	¤	¤

Merging cells in a table

Cells in columns and rows can be merged to improve the appearance of a table. To merge cells:

▶ Highlight the cells to be merged.

▶ From the **Table** menu select the **Merge** menu item.

JULY CONTACTS	
Jennifer	Marton

JULY CONTACTS	
Jennifer	Marton

JULY CONTACTS	
Jennifer	Marton

Sorting tables

Sometimes you will need to sort the text in tables into alphabetical or numerical order. This can be either ascending order (A – Z or 1 – 100) or in descending order (Z – A or 100 – 1).

In the following example we will sort the table into **ascending alphabetical order** of **surnames**:

Surname¤	Initial¤	Title¤	¤
Sunley¤	B¤	Miss¤	¤
Pringle¤	P¤	Miss¤	¤
Benson¤	G¤	Miss¤	¤
Hetherington¤	P¤	Mr¤	¤
Abbotson¤	K¤	Mr¤	¤
Zantos¤	J¤	Mr¤	¤

> **Tip**
>
> You can sort rows in a table by highlighting the ones you want sorted and following this same procedure.

▶ To sort the whole table place the cursor in any cell in the table to be sorted. (It doesn't matter which cell you place the cursor in as Word will automatically select and sort the whole table.)

▶ Select the **Table** menu and then the **Sort** menu item.

The following window is displayed and the whole table will appear selected:

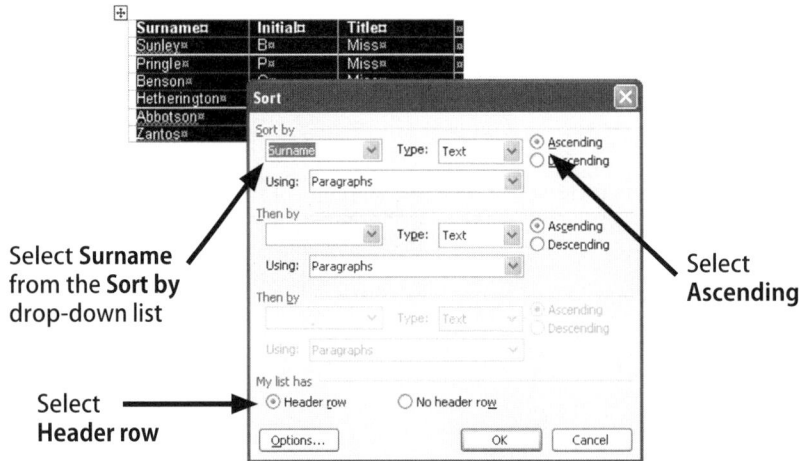

Select **Surname** from the **Sort by** drop-down list

Select **Ascending**

Select **Header row**

- Check that **My list has header row** is selected. This will show the column headings: **Surname**, **Initials**, and **Title** in the **Sort by** drop-down list. Selecting this option will also prevent the column headings being sorted with the other table text and clearly you do not want this in this instance.

- Select **Surname** from the **Sort by** drop-down list. As **Surname** is the first column in this instance this will be selected by default. If you wanted to sort on a different column, click on the drop-down list and select it from the column header row listed.

- Select **Ascending** order.

- Select **OK** and your table will be sorted into ascending alphabetical order of **Surname**.

  Try it and see it work.

Surname	Initial	Title	
Abbotson	K	Mr	
Benson	G	Miss	
Hetherington	P	Mr	
Pringle	P	Miss	
Sunley	B	Miss	
Zantos	J	Mr	

Adding and removing borders to cells

- Highlight the cells to be worked with.

- Select the **Format** menu.

- Select the **Borders and Shading...** menu item.

The window shown opposite appears. This shows a grid that represents the cells you have highlighted.

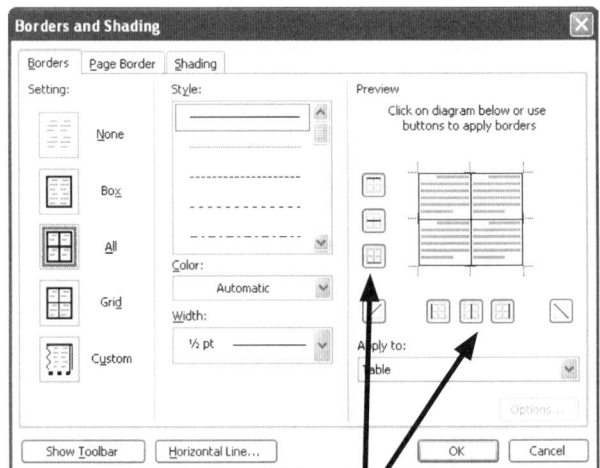

- Click on the buttons around the **Preview** pane to show or hide cell borders.

- Click **OK** when you are happy with your choices.

To change the width of the line or style of the line:

- Click either the **Width** or **Style** required, and then click the line to be changed, or click the boxes around the side of the grid.

- Click **OK** when you are happy with your choices.

Click on any of the buttons around the **Preview** pane to show or hide borders as you wish.

To add shading to the table:

🞂 Highlight the cells to be worked with.

🞂 Select the **Format** menu.

🞂 Select the **Borders and Shading...**
menu item.

🞂 Select the **Shading** tab.

The dialogue box shown opposite will be displayed.

This shows a **Fill** grid and different types of shading. The
Preview area represents the cells you have highlighted.

🞂 Click on the **Fill** required and click **OK**.

To remove shade:

🞂 Click on the **No Fill** box and click **OK**.

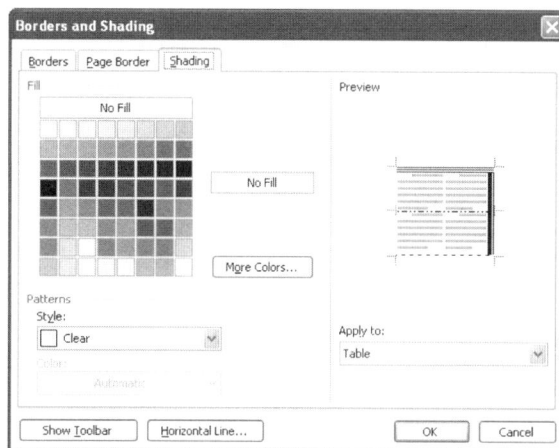

PRACTICE EXERCISE 4 – NUMBERED LISTS AND CREATING A TABLE WITH SHADING

① Open the **Zodiac** document you created for **Practice Exercise 3**.

② Add the following numbered list at the end of the document:

1   Aries
2   Taurus
3   Gemini
4   Cancer
5   Leo
6   Virgo
7   Libra
8   Scorpio
9   Sagittarius
10 Capricorn
11 Aquarius
12 Pisces

> **Tip**
>
> To resize columns, position the pointer
> between the columns. When the cursor
> looks like this ᐊ|ᐅ click and drag the column
> to the size you require.
>
> 23rd Nov
> December

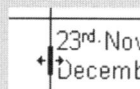

③ Create a table using the information in the table on the following page.

④ Embolden the column headings.

⑤ Adjust column widths, if necessary, to make sure all text appears on one line (*See Tip above*).

⑥ Remove the horizontal borders in the Element column, as indicated.

⑦ Add 15% shading to the rows indicated.

Continued

Practice exercise 4 continued

Element	Zodiac Sign	Relevant Dates
Fire	Aries	21st March – 20th April
	Leo	24th July – 23rd August
	Sagittarius	23rd November – 20th December
Earth	Taurus	21st April – 21st May
	Virgo	24th August – 22nd September
	Capricorn	21st December – 20th January
Air	Gemini	22nd May – 20th June
	Libra	23rd September – 23rd October
	Aquarius	21st January – 19th February
Water	Cancer	21st June – 23rd July
	Scorpio	24th October – 22nd November
	Pisces	20th February – 20th March

8 Print a copy of the document.

9 Change the **numbered list** to a **bullet list** (use any style of bullet you like).

10 Print a copy of the document.

11 Save the document using a different filename.

Drop capitals

▶ Type the first word of your text.

▶ Highlight the first character.

▶ Select the **Format** menu.

▶ Select the **Drop Cap...** menu item.

▶ Select how you want your text to look, and the number of lines of text to drop.

▶ Select **OK**.

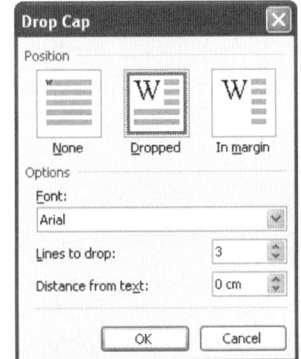

This is an example of a drop capital. It is usually used at the beginning of a document to decoratively highlight the start of it. As you can see, the first character appears larger than the rest and a set number of lines flow around this larger letter. You will often see this used at the start of a chapter in a book or a newspaper article.

Creating columns of text

Sometimes you want your text to have a slightly different appearance, rather like a newspaper column.

This is easy to achieve either before you begin typing, or after you have typed text.

▶ Select the **Format** menu.

▶ Select the **Columns...** menu item.

▶ Select the required number of columns (in the example (*see left*) two columns have been selected).

▶ Tick the **Line Between** check box.

☑ Line between

▶ Select **OK**.

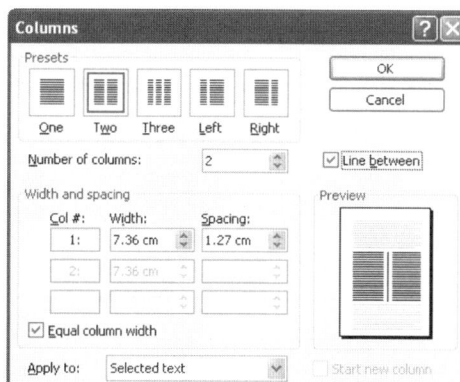

Page breaks and column breaks

Word automatically fills one page or column and then starts another. However, sometimes you want to force text to start a new page or column. To do this:

▶ Position the cursor where you want the page to end.

▶ Select the **Insert** menu.

▶ Select **Break...** menu item.

▶ Select **Page break** or **Column break** from the options displayed in the **Break** window.

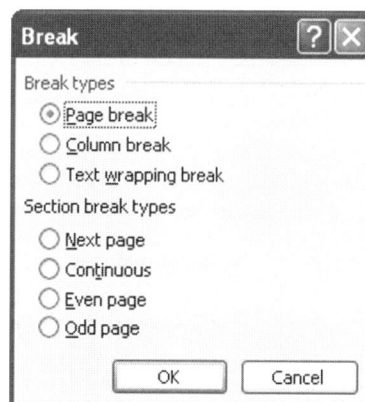

▶ Select **OK**.

Tip: To insert a page break quickly place the cursor where you wish the page to end, hold down the **Control** key and press **Return** key.

PRACTICE EXERCISE 5 – PAGE BREAKS AND SORTING TEXT IN A TABLE

1 Recall the document you saved in **Practice Exercise 4** and amend as follows:

2 After the table, insert a **Page Break**.

3 At the top of the second page, add the following table.

BIRTH SIGNS AND THEIR ASSOCIATED CREATURES AND COLOURS

Sign	Creature	Colours
Pisces	Two Fish	Turquoise, Blues, Sea Green
Sagittarius	The Archer	Fiery Oranges, Purple
Leo	The Lion	Purple, Gold
Taurus	The Bull	White, Blue, Green
Scorpio	The Scorpion	Dark Red, Black
Libra	The Scales	Whites, Pastels
Aquarius	The Water Carrier	Pink, Electric Blue
Virgo	The Virgin	Green, Brown

4 **Centre** and **embolden** the column headings.

5 **Sort** the table on the second page so the **Sign** column entries are in **ascending** alphabetical order.

6 **Print** a copy of the document.

7 **Save** the document with a different filename.

PRACTICE EXERCISE 6 – COLUMNS AND DROP CAPS

① Open Word and create a **New** document.

② Key in the text shown below.

③ Make the heading **Font Tahoma,** and **Font Size 14**.

④ Make the remaining text **Font Size 12**.

⑤ Give the first paragraph a **Drop Cap** of three lines.

⑥ Add **bullets** to the paragraphs where shown.

⑦ Highlight the bullet pointed text and make it into **two columns** with a **line between**.

⑧ **Save** the document using an appropriate filename.

⑨ **Print** a copy of the document.

TRADE DESCRIPTIONS

Are you considering having double glazing, a new conservatory built or refitting your kitchen? Home improvement salesmen are not known for straight talking when it comes to getting us to sign on the dotted line. It helps to be aware of the following:

- Don't fall for the "when will your partner be home?" routine. Some companies think they are more likely to get a sale if both partners are present.
- Don't believe in the "special discount if you sign today". Companies only sell at a price they can afford.
- Don't sign an agreement just to get rid of a persistent salesman. Some believe the longer they are with you the more likely you are to give in and it's surprising how many people do this!

- Salesmen love you to buy on credit because they make commission. If you want to buy on credit, compare the costs with a bank or building society loan.
- Once you have signed a contract it is not always easy to get out of it. If you sign the agreement in your home you are bound by it unless the salesman called uninvited, in which case you have a 7-day cooling-off period, during which time you can cancel.

Working with images

Inserting an image from Clip Art

- Select the **Insert** menu and then the **Picture** menu item, and then, from the pop-out menu select the **Clip Art** menu item.

- Select the image you want to use from the **Clip Art** gallery and Select **Insert** using the down-arrow at the side of the image.

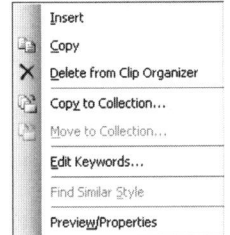

Insert
Copy
Delete from Clip Organizer
Copy to Collection...
Move to Collection...
Edit Keywords...
Find Similar Style
Preview/Properties

Inserting an image from a file

- Select the **Insert** menu and then the **Picture** menu item, and then, from the pop-out menu select the **From File...** menu item.

- Locate where the picture you want to use is stored by using the **Look In** drop-down list.

- Select the picture you want to use and click on the **Insert** button.

Positioning and resizing the image in your document

Text wrapping

Now you will instruct the text to wrap itself around the picture – as you might see in a newspaper.

Picture

In Line With Text
Square
Tight
Behind Text
In Front of Text
Top and Bottom
Through
Edit Wrap Points

- Once an image is in your document click on it. The **Picture Toolbar** should be displayed automatically (if it doesn't, right-click the image and select **Show Picture Toolbar** from the drop-down menu).

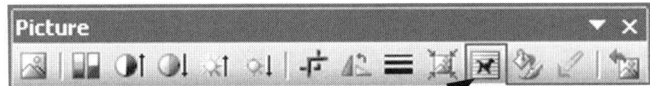

Click on the **Text Wrapping** icon to see the different options

Now you can position your image anywhere in the document and the text will flow around it, giving it a professional look. The practice exercises at the end of this section will help you see how this works.

Resizing and moving images

- Click on the image – if your image has text wrapping set to **In Line With Text** you will notice eight black squares called **handles**. If one of the other options is selected, the handles change to circles and an additional handle is added at the top that can be used to rotate the angle of the image.

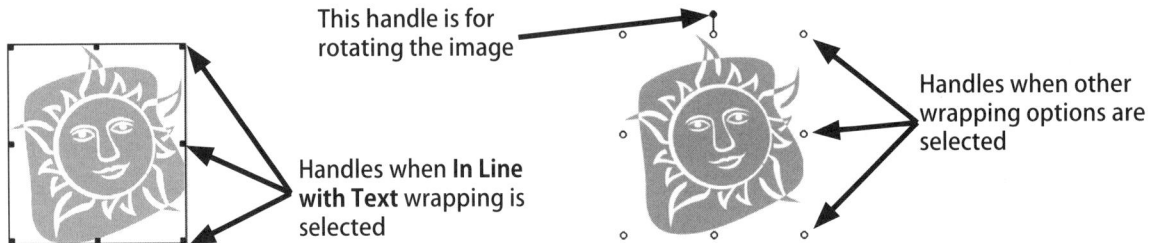

This handle is for rotating the image

Handles when other wrapping options are selected

Handles when **In Line with Text** wrapping is selected

- To resize the image click and drag a corner handle. (If you pick up any of the middle squares on either side, you will change the proportion of the picture. Try it and see.)

- To move the image, click on the image, hold the left-mouse button down and drag the image to the new position.

Copying images

Images can be copied to another place, or page in a Word document, or into another document or even to or from another Microsoft Office application such as Excel. To copy an image:

▶ Highlight the image in the document (e.g. a graph in Excel) so that the handles can be seen (*see page 170*).

▶ Click the **Copy** button on the **Toolbar** (this copies the image to the Windows **clipboard** and it will be lost if you turn off the computer).

From the **clipboard** you can paste the image into another part of the document or another application:

▶ Open the application and document (e.g. a Word document) into which you want the image to be copied.

▶ Place the cursor where you want the image to be placed and click on the **Paste** button.

Headers and footers

The **header** and **footer** of a page form part of the top and bottom margins. On occasions you may need to place information in them. For example, for each Part A Task you will need to enter information such as page numbers, your name and the name of the document in the footer at the bottom of the page.

▶ To access the header and footer, select the **View** menu and then select the **Header and Footer** menu item.

You will then see the toolbar and the header, which has a box with a dotted border, at the top of the page.

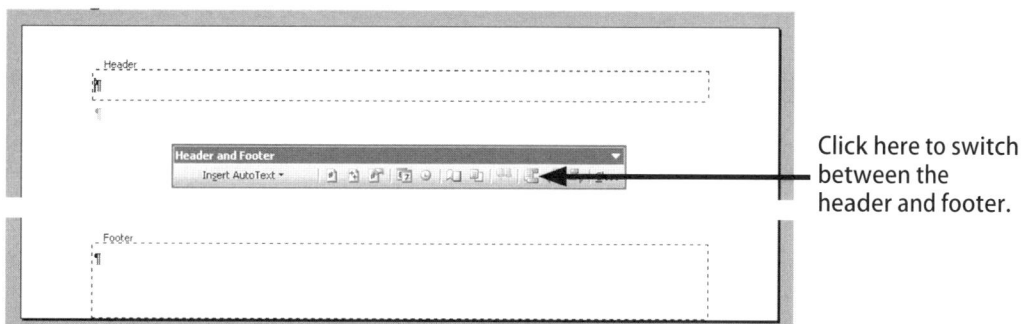

Click here to switch between the header and footer.

You can use the **AutoText** options to insert information such as **filename** or **page numbers**, or you can type these in by hand.

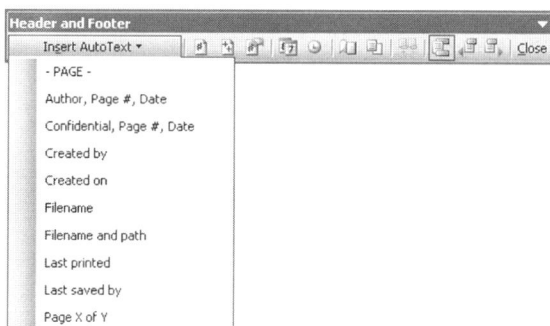

Click on **Close** when completed to close the toolbar.

Your information will appear in the footer. It will look very faint on the screen, but will print out normally.

PRACTICE EXERCISE 7 – ADDING ROWS AND TEXT TO A TABLE, INSERTING AND MOVING CLIP ART

① Open the file you saved in **Practice Exercise 5** and make the following amendments:

② You realise there are four zodiac signs missing from the list on the second page. Add these, inserting rows into the appropriate places so the table remains in alphabetical order.

Capricorn	The Goat	Brown
Gemini	The Twins	Pink, Opalescent
Cancer	The Crab	Silver, White
Aries	The Ram	Red

③ Insert an appropriate image into the Pisces paragraphs on the first page, choosing **In Line With Text** for wrapping of the text around the image.

④ **Print** the document.

⑤ **Delete** the image you inserted into the Pisces paragraph.

⑥ Insert an appropriate image into the Scorpio paragraph, this time selecting **Tight** when wrapping the text around it.

⑦ **Print** a copy of the amended document.

⑧ **Save** the document using a different filename.

Mail Merge

▶ Create a new Word document.

▶ From the **Tools** menu select the **Letters and Mailings** menu item and then **Mail Merge Wizard** from the sub-menu.

The **Mail Merge Task pane** is then displayed on the right-hand side of the screen. There are six steps that you will need to complete in order to create your mail merge document:

Step 1 of 6

▶ Under **Select Document type** on the Task pane select **Letters**.

▶ Select **Next: Starting document**.

Step 1

172

Step 2 of 6

⊙ Under **Select starting document,** select **Use the current document**.

⊙ Select **Next: Select recipients**.

Step 3 of 6

⊙ Under **Select recipients**, select **Type a new list**.

⊙ Under **Type a new list**, select **Create**.

The **New Address List** window is the displayed.

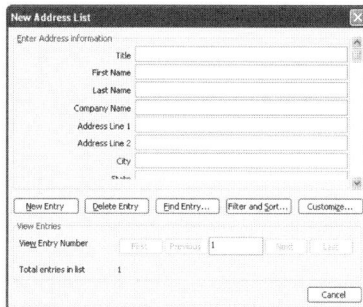

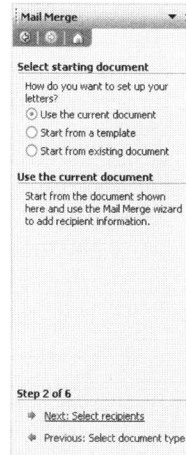

Step 2

You could use the default settings as displayed to create your list of addresses, but we will customise the fields to make adding details easier if you were to ask a friend or colleague to use this at a later time by renaming, deleting, adding and moving them.

⊙ Click on the **Customize** button to display the **Customize Address List** window.

⊙ **Delete** the following fields by selecting them one at a time and then clicking on the **Delete** button. Click **Yes** when you are asked if you are sure you want to delete that field:

Company Name; **Address Line 2**; **City**; **State**; **Country**; **Home Phone**; **Work Phone**; **Email Address**

⊙ **Rename** the following fields by selecting them one at a time and then clicking on the **Rename** button:

First Name as **Initial**; **Last Name** as **Surname**; **Zip Code** as **Postcode**

⊙ **Add** the following fields by clicking on the **Add** button and typing in the name of the field into the **Add Field** window and then clicking **OK**:

Town; **Salutation**; **Time**

Step 3

You can change the order of the fields you created by clicking on a field and clicking on the **Move Up** or **Move Down** buttons. It is not essential to have the fields arranged in the order you will insert them into your letter, but it can save you time when entering information.

⊙ Order the list so that it looks like the *Figure* on the right.

⊙ Click **OK**.

The **New Address List** window now shows the field names you have chosen.

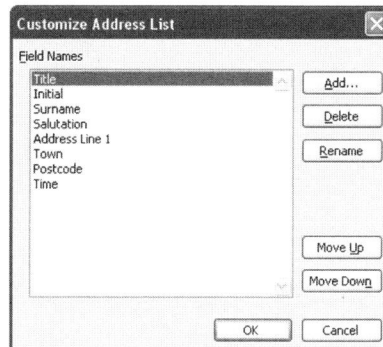

Now enter the information for the six records shown below. Move from field to field by pressing the **Tab Key**. After completing the information for each record select **New Entry**.

Title	Mrs	Dr	The Reverend
Initial	P	S	L
Surname	Salter	Thornton	Cavendish
Salutation	Mrs Salter	Dr Thornton	Reverend Cavendish
Address Line 1	Hollybush Farm	24 Bridge Avenue	The Rectory
Town	Great Ayton	Redcar	Stokesley
Postcode	NR4 2XL	TS29 3TT	NR7 8PW
Time	2.30 pm	2 pm	3.30 pm
Title	Mr	Miss	Miss
Initial	K	B	D
Surname	Everton	Ryan	Keegan
Salutation	Mr Everton	Miss Ryan	Miss Keegan
Address Line 1	51 Hallgarth Mews	10 The Cloisters	6 Edgar Close
Town	Norton	Yarm	Saltburn
Postcode	TS19 4JX	TS26 8EA	NY11 2AU
Time	1 pm	3 pm	1.30 pm

When you have finished entering the records click on the **Close** button.

In the **Save Address List** window, that is then displayed, name the file as **Optician Appointments**. By default the file is saved in the folder called **My Data Sources**, which is usually found in the **My Documents** folder.

Click **Save**.

The **Mail Merge Recipients** window will be displayed. Select **OK**.

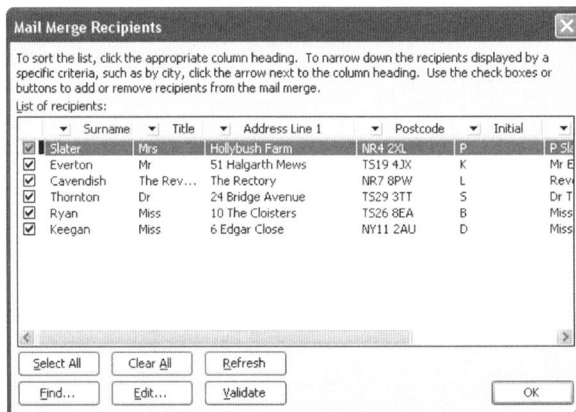

> **Tip**
>
> The letters will be merged in the same order as the **Mail Merge Recipients** list appears. Click on the appropriate column heading to sort the list in ascending or descending order.

Click on **Write your letter** on the **Mail Merge Task pane**.

We are now going to create the letter shown below:

OPTIMAX OPTICIANS

Today's date (dd/mm/yyyy)

≪Title≫ ≪Initial≫ ≪Surname≫
≪Address_Line_1≫
≪Town≫
≪Postcode≫

Dear ≪Salutation≫

Our records show that it is a year since your last eye examination at this practice.

We have set aside an appointment for you to see Mrs Iris Iceton on Tuesday 18th at ≪Time≫.

We would appreciate a confirmation of whether you will be attending this appointment, or alternatively, wish us to make an appointment for you on a different date.

Yours sincerely

I Glass
Practice Manager

▶ Type **OPTIMAX OPTICIANS** and centre the text.

▶ Add a line space after the heading by pressing the **Enter Key** twice when the cursor is at the end of the sentence.

▶ Type today's date and add a line space as above.

Next we will put in the address fields as follows:

▶ Under **Write** your letter select **More items...** to display the **Insert Merge Field** window.

▶ Double-click on each one of the following field names to insert them into the document:

   Title; Initial; Surname; Address Line 1; Town; Postcode

Your document should now look like this so far with the address fields all in one long line and without spaces between the fields:

OPTIMAX OPTICIANS

20/07/2006

«Title»«Initial»«Surname»«Address_Line_1»«Town»«Postcode»

▶ Place the cursor between the fields ≪Title≫ and ≪Initial≫, *see the above Figure* and press the space bar key.

⊙ Repeat this for <<**Initial**>> and <<**Surname**>>.

⊙ Place the cursor between the fields <<**Surname**>> and <<**Address_Line_1**>> and insert a line break by pressing the **Shift Key** and the **Enter Key** at the same time.

⊙ Repeat these steps until the address of your letter looks like the one on *page 175.*

⊙ Leave a line space after the address and type **Dear** and a space and then add in the <<**Salutation**>>field in the same way you added the address fields.

⊙ Type in the remainder of the letter and enter the <<**Time**>> field at the end of the second paragraph.

> ### Remember
>
> Add spaces between field names that are on one line. If you do not do this, when the information is merged it will resemble this example: **ReverendCavendish**

Once your letter is complete it is time to merge it with your **Optician Appointment** data.

⊙ Click on **Preview your letters** on the **Mail Merge Task Pane**.

Step 5 of 6.

⊙ You can preview the individual merged documents by clicking on the **Forward** and **Back** buttons on the **Mail Merge Task Pane**.

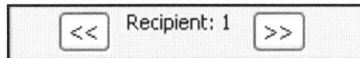

⊙ If you are happy with the way everything looks, click on **Next: Complete the mail merge** at the bottom of the **Task Pane**.

Step 6 of 6.

⊙ Click on **Print...** on the Task Pane to display the **Merge to Printer** window.

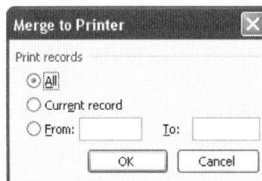

⊙ Select which records to print: **All**, **Current record** or enter values for **From** and **To**.

⊙ Click **OK**.

⊙ **Save** the Mail Merge document and **Exit** from Word.

Step 5

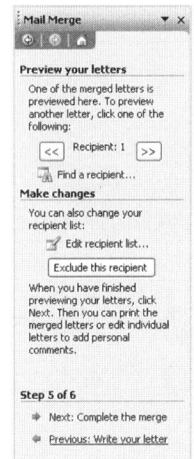

Step 6

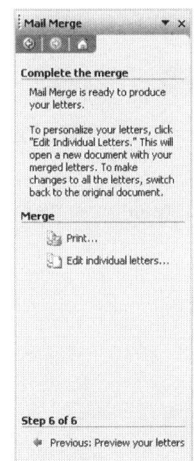

SPREADSHEETS

What is a spreadsheet?

Spreadsheets are widely used in business and are mostly used for analysing figures such as accounts. By adding formulae to a spreadsheet's cells, repetitive tasks can be made much easier. For example, if you were a wine merchant you could create a spreadsheet that worked out the cost of a case of wine based on the unit cost of the wine. If the price of the wine changed you would simply put in the new price and the spreadsheet would automatically work out the case price.

Microsoft Excel is the spreadsheet application used in this book.

Creating a new spreadsheet in Excel

When you first open Microsoft Excel, a blank **Workbook** is normally created by default. A **Workbook** is the name given to an Excel document containing more than one **spreadsheet** (**Worksheet**). If a Workbook has not been created, click on the **New** button.

Title Bar
This displays the name of the document. Unsaved **Workbooks** are called **Book1**, **Book2**, etc., depending how many new **Workbooks** you have created during that session.

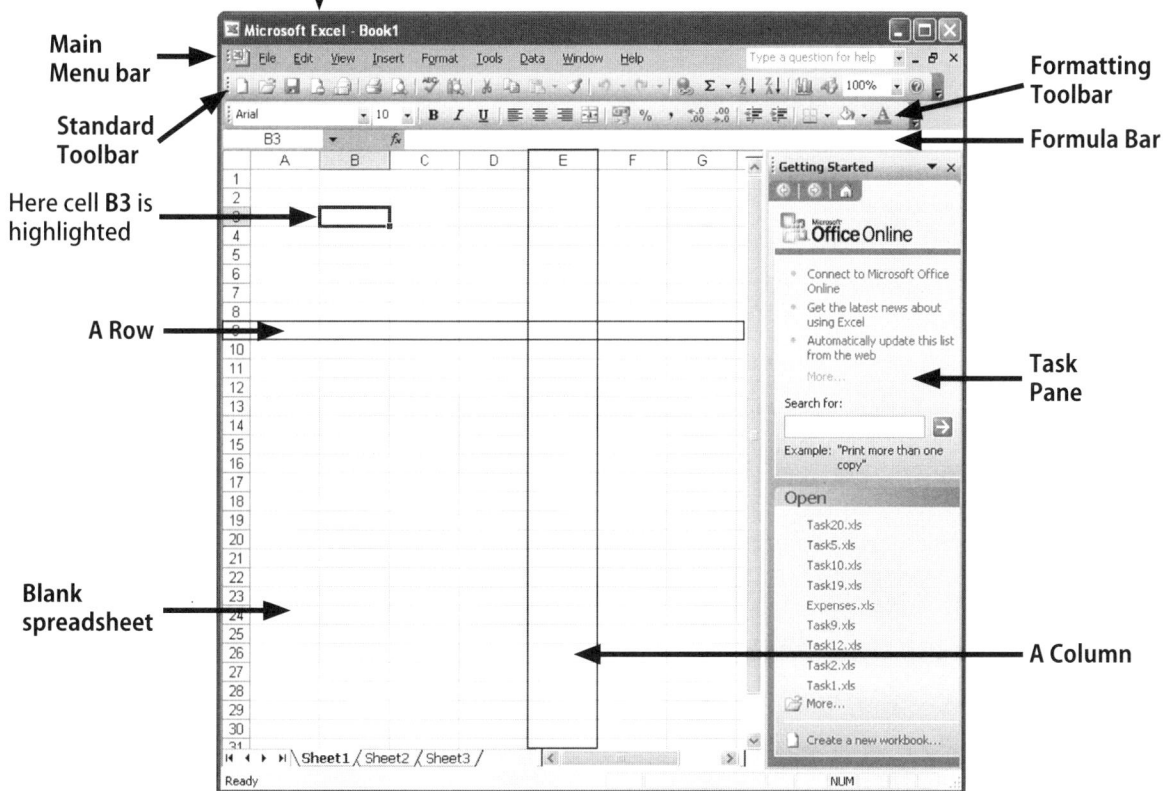

Main Menu bar

Standard Toolbar

Here cell B3 is highlighted

A Row

Blank spreadsheet

Formatting Toolbar

Formula Bar

Task Pane

A Column

A spreadsheet is made up of **columns** and **rows** made up of boxes called **cells**. Each cell has an **address** – like a grid reference on a map. In the figure above, **cell B3** has been selected. You can identify the cell you are working in as it will have a black box around it. You can move around the spreadsheet using the navigation keys on the keyboard.

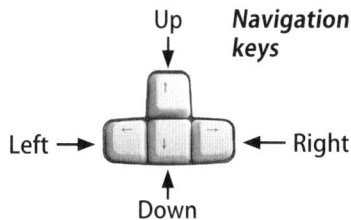

Navigation keys
Up
Left → ← Right
Down

Saving your work

Save

If you intend to keep your work, then you need to **save** it. Do this often as you work to reduce the risk of losing your work. When saving your work for the first time:

➤ Select the **File** menu.

➤ Select the **Save** menu item to display the **Save As** window.

➤ Give your spreadsheet an appropriate filename and choose a folder or drive to save your file to, e.g. a floppy disk or network drive.

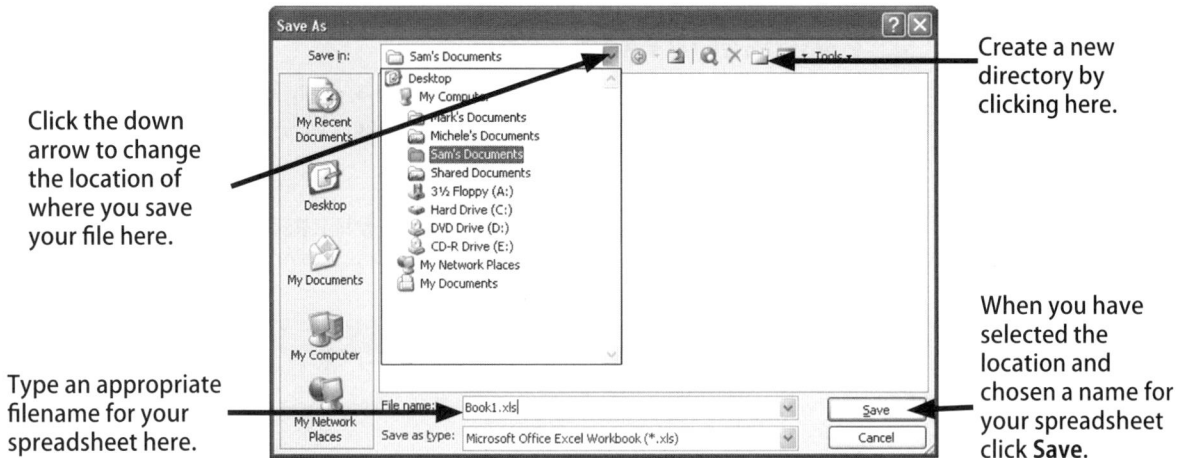

Click the down arrow to change the location of where you save your file here.

Create a new directory by clicking here.

When you have selected the location and chosen a name for your spreadsheet click **Save**.

Type an appropriate filename for your spreadsheet here.

When you save your work again, simply select the **File** menu and **Save** again or click on the **Save** button on the **Standard Toobar**.

The **Save As** window is not displayed in subsequent saves.

Save As

You will often need to save different versions of your documents with new filenames. To do this:

➤ Select the **File** menu.

➤ Select the select the **Save As** menu item to display the **Save As** window.

➤ Give your spreadsheet an appropriate filename and choose a folder or drive to save your file to, e.g. a floppy disk or network drive.

Inserting and editing data

To add data to a spreadsheet:

➤ Click on the cell in which you want to add the data and start typing. You will notice that the text is also displayed in the **Formula Bar**.

A1	▼	*fx*	SANTANA SUMMER SALE		
	A	B	C	D	E
1	SANTANA	SUMMER SALE			
2					

➤ When you have finished typing press the **Enter** key on the keyboard.

Excel assumes that you are entering a column of data and highlights the cell below ready for you to start typing.

To edit the data:

▶ Double-click on the cell you want to edit. This will place the text cursor in the cell ready for you to start editing.

Or

▶ Click on the cell you want to edit and then edit the text displayed in the **Formula Bar**.

Formatting text

To change the **Font** and its **attributes** (**size**, **weigh**t, etc.):

▶ Highlight the cell or cells to be formatted and select one or more of the options available from the **Formatting** toolbar (*see below*).

Font (typeface)

Click on the down arrow to select a font from the drop-down list. Scroll down the list if necessary.

Font attributes

Click on one or more of these buttons to apply the required style. Clicking the button again will remove the formatting.

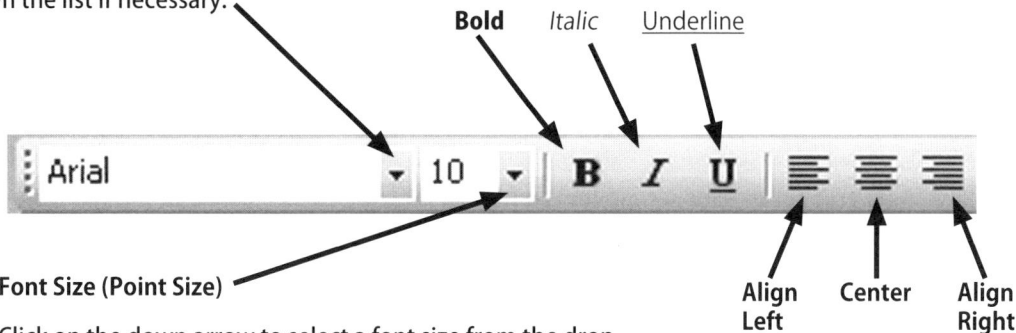

Bold   *Italic*   <u>Underline</u>

Font Size (Point Size)

Click on the down arrow to select a font size from the drop-down list. Scroll down the list if necessary. You can also type a size in if you want to if it's not shown in the list.

Align Left   Center   Align Right

Formatting columns and rows

Changing column widths

The width of a column may be adjusted using one of the following methods:

▶ Double-clicking with the mouse between columns so that the text will fit exactly into the column.

Or

▶ Dragging with the mouse until the desired width is reached.

Or

▶ Highlight column by clicking on the column header (A, B, C, etc.).

▶ Select the **Format** menu and then the **Width** menu item.

▶ Enter the desired width in characters or cms, etc. in the **Column Width** window.

Position the pointer between the columns ✛ and then double-click or drag the column to the size you require.

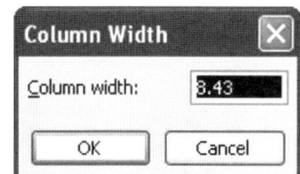

Changing row height

The process is very similar to that for changing column width.

- Highlight the row to be adjusted by clicking on the **row header** (1, 2, 3, etc.).

- Select the **Format** menu and then the **Height** menu item.

- Enter the desired width in characters or cms, etc. in the **Row Height** window.

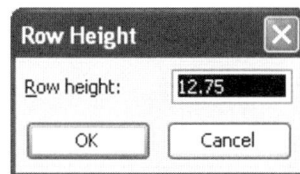

Row Height

Row height: 12.75

OK   Cancel

Inserting a column

- Click on the **column header** (A, B, C, etc.) to the right of where you want to insert the column – this will highlight the entire column.

- Select the **Insert** menu and then the **Column** menu item. This will insert a column to the left of your originally selected column.

Inserting a row

- Click on the **row header** (1, 2, 3, etc.) above where you want to insert the row – this highlights the entire row.

- Select the **Insert** menu and then the **Rows** menu item. This will insert a row below the one you originally selected.

Deleting a column

- Highlight column by clicking on the **column header** (A, B, C, etc.).

- Select the **Edit** menu and then the **Delete** menu item.

Deleting a row

- Highlight the row to be deleted by clicking on the **row header** (1, 2, 3, etc.).

- Select the **Edit** menu and then the **Delete** menu item.

Hiding a column or row

- Highlight column by clicking on the **column header** (A, B, C, etc.).

- Select the **Format** menu and then select the **Column** menu item and then the **Hide** menu item.

This process can also be used to hide rows. You will need to select the row then choose **Format**, **Row**, **Hide**.

Unhiding a column or row

- Highlight the columns either side of the hidden column by clicking and dragging on the **column headers** (A, B, C, etc.).

- Select the **Format** menu and then select the **Column** menu item and then the **Unhide** menu item.

This process can also be used to unhide rows. You will need to select the rows either side of the hidden row and then choose **Format**, **Row**, **Unhide**.

Displaying numbers in integer format (no decimal places)

Integer format means whole numbers.

> Select the cell (or range of cells) to be displayed in integer format.

> Select the **Format** menu and then select the **Cells** menu item.

> In the **Format Cells** window select the **Number** tab.

> Under the **Category** heading select **Number** and reduce the **Decimal places** to **0**.

Excel will round up or down, e.g. 2.36 will become 2 or 3.78 will become 4.

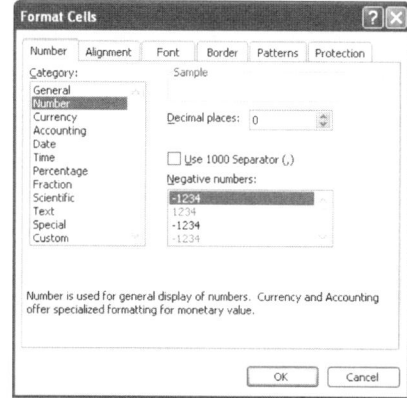

Displaying numbers to two decimal places

> Repeat the procedure above, but select number of decimal places as **2**.

If this is not done, the display of numerical data will appear uneven. For example, 2.50 will be displayed as 2.5, whereas 2.39 will be displayed as such.

Displaying pounds and pence

> Select one or more cells, rows or columns.

> Display the **Format Cells** window as you did for displaying numbers above.

> Select the **Number** tab and click on **Currency** under the **Category** heading. Then select £ and click **OK**.

Displaying percentages

> Select one or more cells, rows or columns.

> Display the **Format Cells** window as you did for displaying numbers above.

> Select the **Number** tab and click on **Percentage** under the **Category** heading and click **OK**.

See also page 185.

Wrapping text

Sometimes the text will not fit into a cell and needs to wrap around onto the next line, but not creating another row.

To do this:

> Select the cell that needs to have the text wrapped.

> Select the **Format** menu.

> In the **Format Cells** window select the **Alignment** tab.

> Select the **Wrap text** box by clicking once with the mouse. A cross will appear in the checkbox.

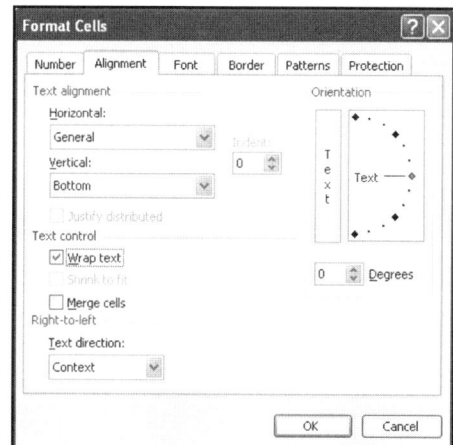

> Click **OK**.

Merging cells in a spreadsheet

If you enter a heading into a spreadsheet it may not fit exactly into one cell, but span over several cells. The example, right, shows a heading entered in cell **A1** before merging the cells.

When the remainder of the spreadsheet is constructed, this heading may look out of place and so cells can be merged to improve its layout.

	A	B	C	D	E
1	The Excelsior Café				
2		03/07/2006	10/07/2006	17/07/2006	24/07/2006
3	Toast	25	35		
4	Bread Rolls	20	40		
5	Soup	50	15		
6	Sandwiches	35	55		

To merge the cells for the heading:

▶ Highlight cells **A to E** in **row 1** then use the **Merge** button on the **Standard Toolbar**.

The result will look like this:

	A	B	C	D	E
1	The Excelsior Café				
2		03/07/2006	10/07/2006	17/07/2006	24/07/2006
3	Toast	25	35		
4	Bread Rolls	20	40		
5	Soup	50	15		
6	Sandwiches	35	55		

Rotating and aligning text

Sometimes you will want to change the direction in which text appears in a cell. This may be to enhance the spreadsheet's appearance or for a practical reason of fitting more text onto a page. In the Excelsior example above the dates take up a great deal of space and if we had a whole month's worth of figures it may not fit on page.

▶ Highlight the cells to be changed. In this instance it is **B2 – E2**.

▶ Select the **Format** menu and then the **Cells** menu item.

▶ Select the **Alignment** tab.

▶ In the **Degrees** number box, increase the number to 65. Look at the **Orientation** preview to see how the text will look.

▶ Ensure that the **Text Alignment Horizontal** is set to **General** and that the **Vertical** is set to **Bottom**.

▶ Click **OK**.

▶ Adjust the column widths of cells **B2 – E2**

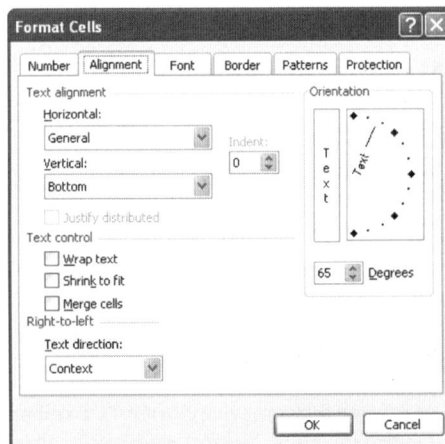

The result will look like this:

	A	B	C	D	E	F
1	The Excelsior Café					
2		03/07/2006	10/07/2006	17/07/2006	24/07/2006	
3	Toast	25	35			
4	Bread Rolls	20	40			
5	Soup	50	15			
6	Sandwiches	35	55			

182

Adding borders to cells

Text and numbers can be difficult for the reader to understand unless there is ruling around and between the columns and rows.

- Highlight the columns and rows you want to be given borders.

- Select the **Format** menu and then the **Cells** menu item. This will display the **Format Cells** window (*see right*).

- Select the **Borders** tab.

- Select **Outline** and **Inside** to place borders around all sides of each of the highlighted cells.

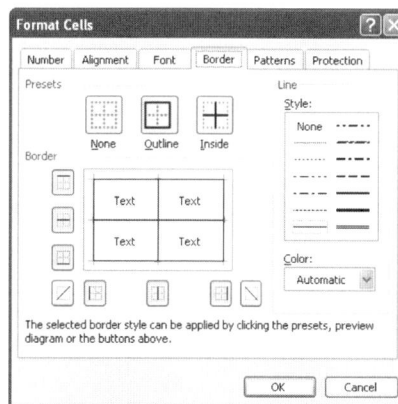

- Select **OK**.

Adding shading to cells

Adding shading to tabular information can, sometimes, make data easier to read and understand at-a-glance. To do this:

- Highlight the cells you want to apply the shading to.

- Select the **Format** menu and then the **Cells** menu item. This will display the **Format Cells** window.

Choose the main colour for your selected cells here.

Click on the drop-down arrow to choose a pattern...

...and the pattern colouring here.

A preview of how the shading and pattern might look is shown here.

- Select the **Patterns** tab.

- Select the colour you require from the **Cell Shading Color** pallet.

- If you want to apply a pattern as well click on the drop-down **Pattern** menu and make your colour selection and a colour for that pattern if you wish.

- Select **OK** when you are happy with your choices.

3

Formulae

One of the main reasons for using spreadsheets is to analyse information and data. Using a spreadsheet's powerful formulae you can quickly calculate values entered in the columns or rows of data. Formulae can be typed directly into the **Formula Bar** simply by putting an equals sign (=) before each formula.

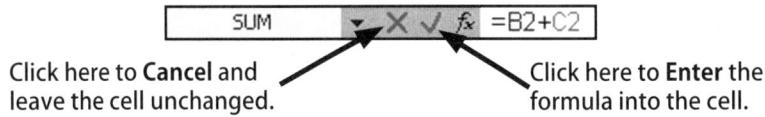

| SUM | ▼ | ✕ | ✓ | *fx* | =B2+C2 |

Click here to **Cancel** and leave the cell unchanged.

Click here to **Enter** the formula into the cell.

In the following examples we will look at addition, subtraction, division and multiplication. We will put values into cells **B2** and **C2** and the results will be calculated by putting a formula in cell **D2**.

Addition (+)

D2	▼		*fx*	=B2+C2		Formula Bar
	A	B	C	D		
1						
2		2	3	5		Cell D2

Click on **cell D2** and type the formula in the **Formula Bar**. Don't forget the **=** sign at the start!

Subtraction (-)

D2	▼		*fx*	=B2-C2
	A	B	C	D
1				
2		5	2	3

Division (/)

D2	▼		*fx*	=B2/C2
	A	B	C	D
1				
2		10	5	2

Multiplication (*)

D2	▼		*fx*	=B2*C2
	A	B	C	D
1				
2		2	6	12

Adding a range of cells

The **Sum** formula can be used to quickly total a range of numbers such as a column or row without having to use a plus (+) or minus sign (-) between each cell reference. To find the total of a row of numbers do the following:

▶ Click in a cell were you want the result to be displayed (in the example below we have chosen **H4**).

▶ In the **Formula Bar** type in =SUM(

▶ Next, click on the first cell in range to be added (in this instance **B4**), hold the mouse button down and drag to include all cells to be added together (in this instance we have gone to cell **G4**.

▶ Go back to the **Formula Bar** and type in a closing bracket).

The colon (:) between the two cell references means **B4 and everything up to and including G4**, i.e. the colon (:) takes the place of the words **and everything up to and including**.

SUM	▼	✕ ✓	*fx*	=SUM(B4:G4)			
	A	B	C	D	E	F	G	H
1								
2								
3								
4		20	15	16	220	67	29	(B4:G4)

▶ Click on the **Enter** button. ✓

Copying formulae (replicating cells)

It is useful to be able to copy a formula to another cell, for example if you wanted to total up a number of individual rows. To do this:

➤ Select the cell containing formula so that it has a black border. You will notice a small black box (a handle) in the bottom-right corner.

Click on the black handle

```
367
```

➤ Click on the hold the mouse button down on the handle and drag it over cells where you want to copy formula into. This can be done across rows as well as up or down columns.

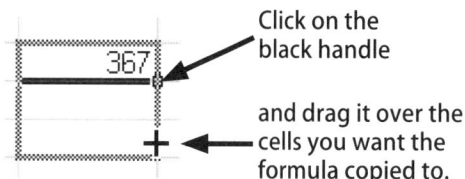

and drag it over the cells you want the formula copied to.

Percentage formula

You will often find that you will be asked to calculate percentages, for example your firm may plan to increase the price of its goods by 10%. By creating a spreadsheet this task can become very easy once you know the formula.

We will now create a spreadsheet that can work out percentage increases or decreases.

➤ **Open** Excel and create a **New Blank workbook**.

➤ Enter the headings, **embolden** and **wrap** the text as shown in the figure on the right.

	A	B	C
1	Price	% Increase or Decrease	New Price

➤ Highlight **cells A2 – A5** and format them as **Currency (£)** by clicking on the **Currency** button on the **Main toolbar**.

➤ Highlight **cells B2 – B5** and format them as **Percentage** by clicking on the **Percentage** button on the **Main toolbar**.

Note

Format the cells with **Percent** BEFORE entering figures as Excel multiplies the figure in the cell by 100 to obtain the percentage. Therefore if you had entered **10** and then applied the Percent format, the figure shown would be **100**. By applying the formatting first, Excel will display the figure as **10%**, but store the number as **0.1**.

➤ By default, Excel will display percentages rounded to the nearest integer, but Excel does store and use the figure as you entered it. To make this exercise easier to read highlight **cells B2 – B5** and format them to **2 decimal places**, *see page 181 on details how to do this,* or alternatively by clicking twice on the **Increase Decimal** button on the **Main toolbar**.

➤ Enter the following formula in **cell C2**:

=A2*B2+A2

The first part of this formula, **A2*B2**, works out what the value of the percentage is of the price by multiplying the **Price** by the **Percentage** (e.g. **£100 x 0.1 = £10**). The result of this (**£10**) is then added to the original **Price** in **A2** and gives us the total increased **Price** of **£110.00**.

Remember: 0.1 is the number Excel stores as the value for 10%. Another way to consider this formula is =A2*0.1+A2.

➤ Copy/replicate the formula in **cell C2** to **cells C3 – C5**.

▶ Enter the figures shown in the *Figure* below in **A2 – A5** and **B2 – B5** and see the results you obtain in **column C**. To see the effects of a decrease in percentage simply place a minus sign (-) in front of the percentage amount, *see **cell B3*** in the figure below:

	A	B	C
1	Price	% Increase or Decrease	New Price
2	£ 100.00	10.00%	£ 110.00
3	£ 100.00	-10.00%	£ 90.00
4	£ 88.00	20.00%	£ 105.60
5	£ 97.56	23.70%	£ 120.68

Do not type the numbers shown in column C. These results will be calculated by the formula that you entered earlier.

Another way of finding the percentage value of a figure is to multiply that figure by the percentage value and include the **%** sign in the calculation, for example to work out the VAT value of a cell, use the formula:

$$=\textbf{cell ref*17.5\%} \quad (\text{e.g.} =\text{A2*17.5\%})$$

The Function Wizard

We have looked at how to type in formulae into the **Function Bar**, but Excel also has a **Function Wizard** that helps you write functions. It can be used to help create any of the built in functions such as **sum**, **average**, **date**, **count**, **max** and **min**.

▶ Open the **Function Wizard** by clicking on the **Insert Function** button, which is positioned next to the **Formula Bar**. f_x

The **Insert Function** window will be displayed.

Type in a word or phrase, click on the **Go** button and the wizard will provide a list of possible functions to choose from.

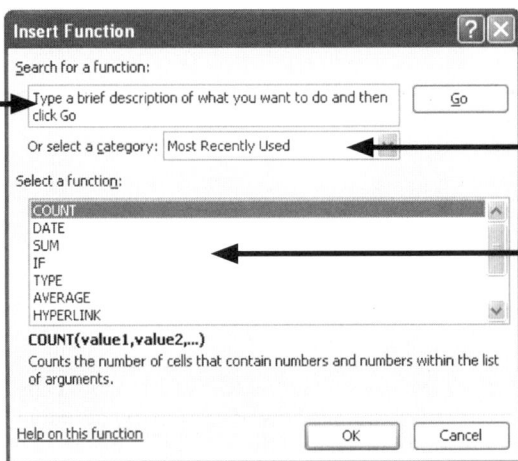

```
Insert Function                          [?][X]
Search for a function:
[Type a brief description of what you want to do and then    Go
click Go]
Or select a category: Most Recently Used ▼
Select a function:
COUNT
DATE
SUM
IF
TYPE
AVERAGE
HYPERLINK
COUNT(value1,value2,...)
Counts the number of cells that contain numbers and numbers within the list
of arguments.
Help on this function          OK        Cancel
```

Click here to see a list of categories for all of the functions available.

The list of functions displayed here depends upon the category selected.

Select the function you require and then click **OK**.

▶ Select the function you require and click on the **OK** button. This will display the **Function Arguments** window.

Average formula example

▶ Type some numbers in a column and select the next empty cell below (*see Figure right*).

▶ Open the **Function Wizard** and locate the **Average** function using one of the methods shown in the figure above and select **OK**.

25
70
33
47

The **Function Arguments** window is displayed. Notice that the wizard has guessed the range of cells for the numbers you want to work out the average for. If these are incorrect, click on the first cell you want to include and drag the mouse over the remaining cells to be included.

The wizard guesses the range. Change this if necessary.

You can preview the result.

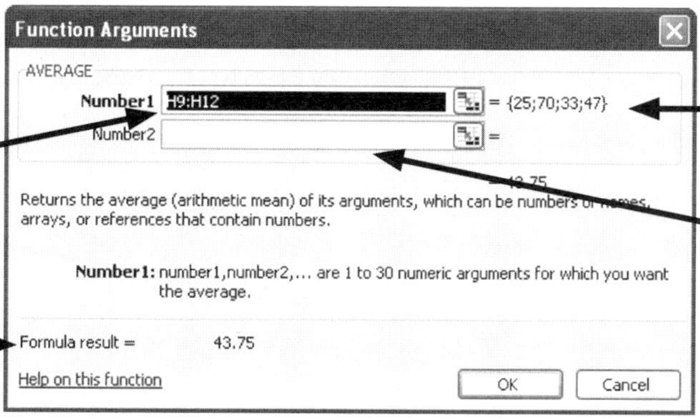

The values of each cell are shown here.

More than one column or row can be added to the formula if necessary.

Click **OK** and the average figure will be show in the cell.

Headers and footers

Headers and footers can be used to provide extra information about the spreadsheets that can appear on each page. For your Key Skills portfolio it is often necessary to prove you have been working on it over a period of time and you may need to add in your name, current date and filename to your documents. To add in this information do the following:

Select the **View** menu and then the **Header and Footer...** menu item to display the **Page Setup** window.

Note

The date and filename appear as codes and will be updated automatically as changes are made.

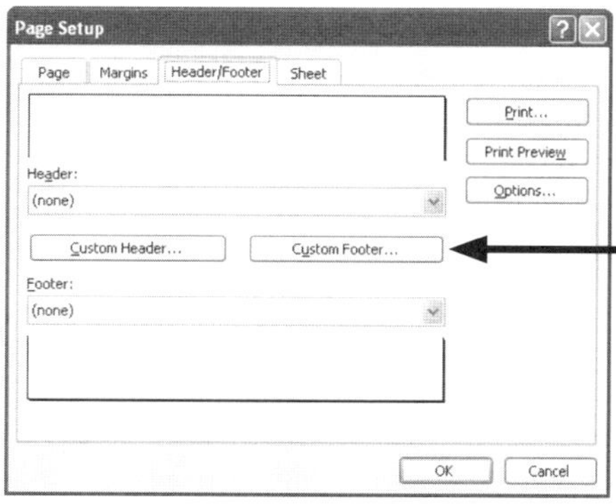

Click on **Custom Footer.**

Click on the **Custom Footer...** button and follow the instruction on the following figure:

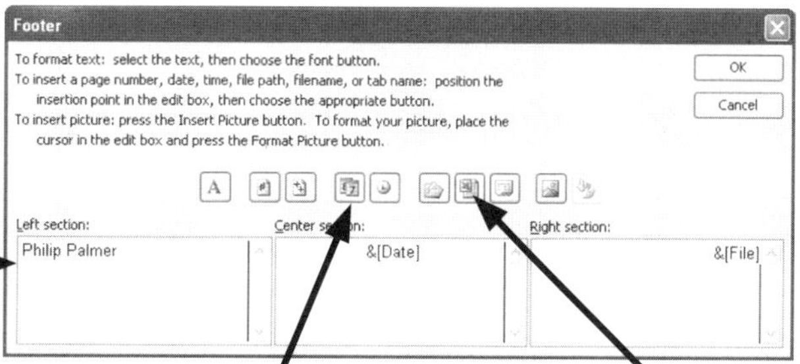

1 Type in your name here.

2 Click in the **Centre section** box and then click on the **Date** button to insert the current date.

3 Click in the **Right section** box and then click the **Filename** button to insert the name of your document.

187

Information and Communication Technology: Reference Sheets 3

Printing a spreadsheet

◗ To print, either select the **File** menu and then the **Print** menu item.

◗ Or just simply click on the **Print** button on the **Standard Toolbar**.

If the spreadsheet is large, either by width or height, you may need to adjust the print settings.

In **Page Setup** you have a number of options you can choose to make the spreadsheet fit on a page. You can:

✔ use **Scaling** to reduce the size or enlarge as appropriate;

Tip: As formulae are often very long, the information may not always fit onto one page, even when printed in landscape. Always **Select Fit to 1 page(s) wide**.

✔ change the orientation of the page from **Portrait** to **Landscape**;

✔ scale the page to fit, say, one page wide by one page tall.

Printing out formulae

If is very useful and a requirement for your Key Skills portfolio to be able to print out a version showing the formulae of your spreadsheet.

◗ Before you can do this you must change the print settings by selecting the **Tools** menu and then the **Options...** menu item.

◗ In the **Options** window select the **Formulas** box as shown in the figure and click **OK**.

◗ Next select the **File** menu and then the **Page Setup...** menu item.

Select the **Formulas** check box.

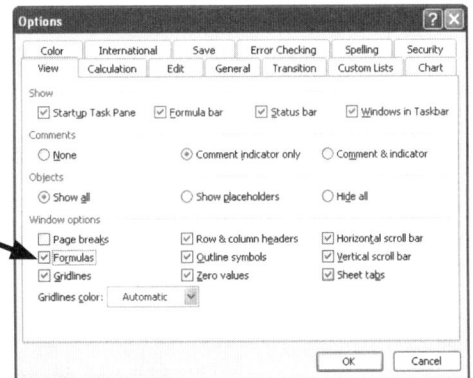

◗ In the **Page Setup** window select the **Page** tab if it is not already displayed and change the **Orientation** to **Landscape**.

◗ Change the **Scale** option to **Fit to 1 page(s) wide** and click **OK**.

◗ Click on the **Print Preview** button on the **Standard toolbar** to check that all the formulae will be seen when printed.

This is the view that will be printed out:

	A	B	C	D	E	F
1			The Excelsior Café			
2		38901	=B2+7	=C2+7	=D2+7	Total
3	Toast	25	35			=SUM(B3:E3)
4	Bread Rolls	20	40			=SUM(B4:E4)
5	Soup	50	15			=SUM(B5:E5)
6	Sandwiches	35	55			=SUM(B6:E6)

PRACTICE EXERCISE 8 – CREATING A SPREADSHEET

1   Create a **New blank workbook**.

2   Enter the information shown in the *Figure* below, beginning in **cell A1** with the heading.

	A	B	C	D	E
1	FEATHERED FRIENDS FEEDING GUIDE PRICE LIST				
2	Code	Description	Cost	No. in Stock	Stock Value
3	A446	Mealworm Food 250g	£2.00	456	
4	A447	Mealworm Food 900g	£5.10	530	
5	A817	Dried Mealworms 100g	£6.95	157	
6	B818	Dried Mealworms 200g	£12.95	204	
7	B485	Feedsafe Feeder	£18.95	166	
8	C189	Fruit & Nut Treat	£2.50	278	
9	C190	Cherry Treat	£2.50	185	
10	C191	Pepper Treat	£2.50	305	
11	C192	Apple Treat	£2.50	135	
12	C193	Insect Treat	£3.00	181	
13	D113	4-season Feeder Mix 500g	£2.60	672	
14	D115	Softbill Mix 4kg	£7.90	298	
15	**TOTALS**				

3   Adjust the cell widths so that all text can be seen. Do this as often as necessary when creating this spreadsheet.

4   Where necessary, **wrap** column headings so the heading is not greatly longer than the column widths.

5   Format **columns C** and **E** as **currency** with the **£** sign and to **2 decimal places**.

6   Enter a formula in **cell E3** of the **Stock Value** column to calculate the total value for that stock item.

7   Copy the formula to **cells E4 – E14**.

8   Insert a formula to provide totals for the **No. in Stock** and **Stock Value** columns.

9   Format all of the cells with **borders**, but do not include the heading in **row 1**.

10   **Merge** the cells in **row 1** so that the heading flows across the width **cells A – E** and centre the heading.

11   **Save** the spreadsheet with a suitable filename.

12   Insert a **Custom Footer** in the spreadsheet that will show the following:

Your name   *The spreadsheet filename*   *Today's date*

13   Print the spreadsheet twice: once showing the totals and once showing the formulae. Adjust to fit on one page (landscape) and make sure all entries are shown in full.

14   **Save** the spreadsheet and exit from Excel.

Information and Communication Technology: Reference Sheets

3

Sorting data

A powerful feature of spreadsheet programs is their ability to sort data in alphabetical or numerical order. However, you need to take care when sorting data as Excel will, if you let it, sort just one column of data and can therefore change the values of your data. In the example we have used below, if we were to sort the prices in **column C**, in the *Figure* below, in descending order the **Embroidered Shorts** would become priced at £9.99.

To carry out a sort and keep all the accompanying data together:

▶ Select all the cells to be sorted. In the example below, **cells A2** to **E17** have been selected. Note that the column headers have been included in the selection.

	A	B	C	D	E
1		SANTANA SUMMER SALE: LADIES GOODS			
2	Item Code	Description	Price	No in Stock	Stock Value
3	A446	Embroidered Shorts	£4.50	497	£2,236.50
4	A447	Embroidered Shirt	£9.99	471	£4,705.29
5	B489	Tie Belt Cotton Trousers	£5.00	381	£1,905.00
6	C176	Printed T-Shirts - Red	£3.00	802	£2,406.00
7	C177	Printed T-Shirts - Grey	£3.00	509	£1,527.00
8	C178	Printed T-Shirts - White	£2.50	109	£272.50
9	B916	Tiered Peasant Skirt	£6.00	143	£858.00
10	B599	Sequined Camisole - Black	£8.50	158	£1,343.00
11	B560	Sequined Camisole - Red	£8.50	206	£1,751.00
12	C866	Lace Trimmed Cotton Blouse	£2.00	200	£400.00
13	C339	Rainbow Tie Shorts	£2.50	200	£500.00
14	C340	Black Tie Shorts	£2.00	197	£394.00
15	B917	Halter-neck Cotton Print Top	£2.95	80	£236.00
16	B335	Beaded Camisole - Red	£5.00	135	£675.00
17	B336	Beaded Camisole - Silver	£4.50	76	£342.00
18	TOTALS			4,164	£19,551.29

▶ Select the **Data** menu and then the **Sort...** menu item. This will display the **Sort** window.

▶ Ensure that **Header row** option selected.

▶ Select the name of the column you want to sort in the Sort by option list. In the example, **Price** has been selected.

▶ Select either an **Ascending** or **Descending** option.

▶ Click **OK**.

> **Sort Ascending** and
>
> **Sort Descending** buttons on the **Standard Toolbar** are a quick way of sorting data. When you use these the **Sort Warning** window appears. Select the **Expand the selection** option to ensure all of the data is sorted together.

Sort

Sort by
Price — Ascending / Descending

Then by
— Ascending / Descending

Then by
— Ascending / Descending

My data range has
⦿ Header row ○ No header row

Options... OK Cancel

PRACTICE EXERCISE 9 – MORE FORMULAE, WRAPPING TEXT AND SORTING DATA

① Open the spreadsheet you saved in **Practice Exercise 8** and make the following amendments:

② Insert columns for **No. Sold** and **Remaining Stock Levels** to the right of the final column **Stock Value**.

③ Adjust the heading **Remaining Stock Levels** so the heading is **wrapped** and **aligned vertically** from the top. It may be necessary to increase the height of **row 2**.

④ Format these columns as **integer**.

⑤ Adjust the appearance of the last five columns so the figures are **left aligned**.

⑥ Insert the following figures in the **No. Sold** column. (**DO NOT ADD THE CODE DATA AGAIN** – we have shown the **Code** column in order to help you match the correct quantity to the product code.)

Code	No. Sold
A446	205
A447	349
A817	47
B818	97
B485	82
C189	73
C190	103
C191	196
C192	88
C193	84
D113	301
D115	106

⑦ Insert a formula in **cell G3** that will calculate the figure which represents the **Remaining Stock Levels** (i.e. **No. in stock** minus **No. sold**).

⑧ Copy the formula to **cells G4 – G14**.

⑨ **Sort** the spreadsheet so the **Cost** column is in **ascending** price order.

> **REMEMBER** to check that **My data range has Header row** is selected in the **Sort** window so Excel knows not to include this in the sort.

⑩ Add a formula in **cell C15** that will give you the total of that column.

⑪ Extend the borders to include the new columns.

⑫ Enter formulae in **cells D15, F15 and G15** to total those columns.

Continued

13 Adjust the heading row so that it is **merged** and **centred** across the width of all the data and appears in **Italics**.

14 **Save** the workbook with a new filename **Feathered Friends 2**. Check that the footer shows today's date, your name and that the new filename is displayed.

15 **Print** the spreadsheet **twice**: once showing the formulae, and once showing the totals. Make sure all entries are shown in full.

Make the following additional amendments:

16 Change the **column heading** in **D2** to read **Number in Stock as at 30 October**.

17 **Wrap** the text so the heading fits into the column's width.

> **NOTE:** It is not good practice to have a heading that is very much wider than the width of the column beneath it. If this occurs then **wrapping the text** is advised.

18 Add a final column at the right of the spreadsheet with the heading **Value of Items Sold**, and **wrap** the text as in *step 17* above.

19 **Format** the column as **Currency** and to **2 decimal places**.

20 Add a formula that will calculate the **Value of the Items Sold** in **cell H3** and copy it to **cells H4 – H14**.

21 Enter a formula in **cell H15** to calculate the **Total Value of the items sold**.

> **NOTE:** If you see the ######## entry in E15 it is because the cell is not wide enough to display the **Total**. Increase the column width until the figures are shown (*see page 179*).

22 Add a formula in **cell E3** that will calculate the **Stock Value (Cost x No. in Stock)**

23 Replicate the formula in **cell E3** down to **cell E14**.

24 Add a formula to calculate the **Total** of **column E**.

25 Add **borders** to **column H** to match the rest of the spreadsheet and readjust the main heading so it is **merged** across **column H** as well.

26 **Save** the workbook with a new filename **Feathered Friends 3**. Check that the footer shows today's date, your name and that the new filename is displayed.

27 **Print** the spreadsheet **twice**, once showing the formulae and once the totals. **Remember** each printout must fit on one page and all entries must be visible.

28 **Save** the spreadsheet.

Adding images from Clip Art

This is same procedure as used for Word (*see page 170*). When you add an image on a spreadsheet you may need adjust the height or width of cells to improve the layout. See *pages 179 and 180* on how to do this.

Creating graphs and charts

An Excel chart is a graphic or diagram based on the numbers, text and calculations that are located in the rows and columns of an Excel worksheet.

Creating a chart using the Chart Wizard

- ▶ Type the following data into a new spreadsheet document:

	A	B
1	Visitors to Museum	
2	Year	Visitor Numbers
3	2003	1950
4	2004	1750
5	2005	1650

- ▶ Select the cells that contain the data you want to make into a chart (**B3 – B5**):

	A	B
1	Visitors to Museum	
2	Year	Visitor Numbers
3	2003	1950
4	2004	1750
5	2005	1650

- ▶ Select and run Excel's **Chart Wizard**.

- ▶ Work through the four steps in the **Chart Wizard** window as follows:

- ▶ **Step 1 – Chart Type:** Select the **Chart type** you want. Here we will select a **Column** type.

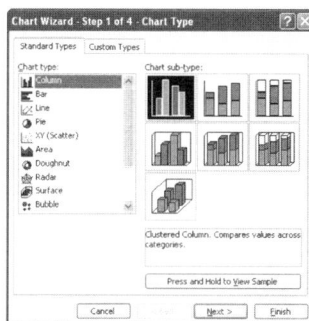

- ▶ Click **Next**.

- ▶ **Step 2 – Chart Source Data:** Select the **Series** tab and then click inside the **Name** box so that the cursor is flashing. Leave the **Wizard** open and select **cell A1**. This will add the text from that cell as the **Series** name.

  Click in the **Category X axis** box on the same screen of the Wizard so that the cursor is flashing. Leave the **Wizard** open and select **cells A3 – A5**. This will add the **Years** from those cells to the **X axis**.

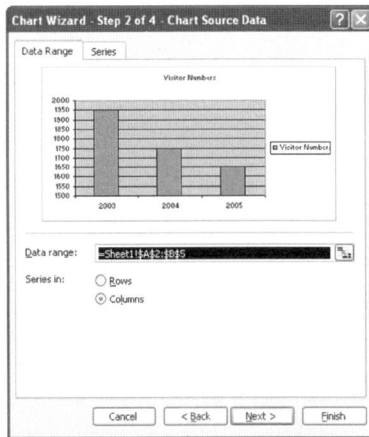

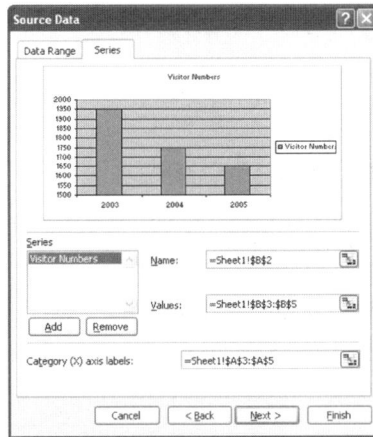

◉ Click **Next**.

◉ **Step 3 – Chart Options:** Change the chart title to **Visitors to Museums** in the **Chart title** box. Type **Year** into **Category X axis** and type **Number of Visitors** into the **Value (Y) axis** box.

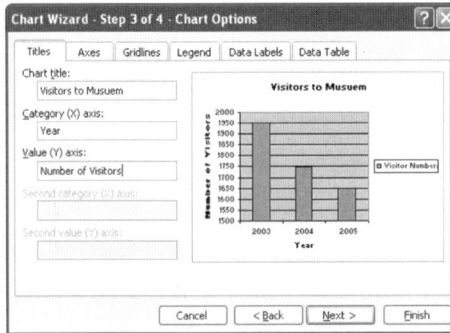

◉ Click **Next**.

You are now asked if you want to place the chart **As new sheet: Chart1**, which will create a new sheet for the chart, or **As object in: Sheet**, which will place the chart on the current worksheet.

◉ Select **As object in: Sheet1**.

◉ Click **Finish**.

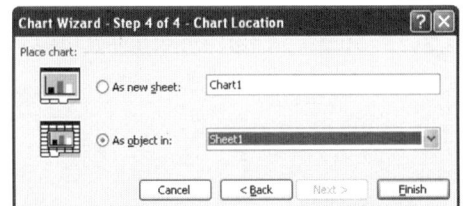

The chart is placed on the current worksheet.

The final chart

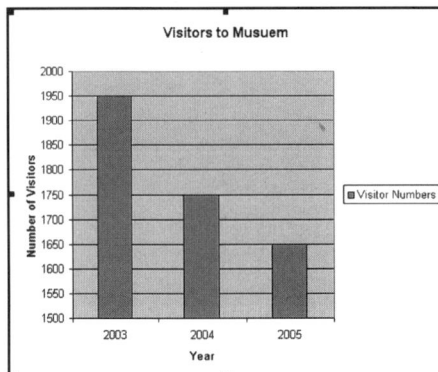

194

Moving a chart on a worksheet

You can **move** the chart to a different position by clicking and holding down the left-mouse button on any blank part of the **Chart Area** (i.e. not on any labels or graphics) and then dragging the chart to a new position.

Data points and data series

Excel charts are based on two basic ideas: the **data point** and the **data series**. A data point has two parts: the item and the value. A data series is a collection of data points.

When Excel draws a chart with more than a single data series, it uses a different colour to represent each series.

Chart types

There are four commonly used chart types:

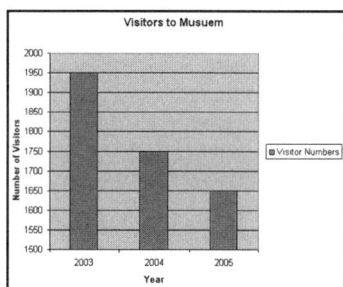

Column Charts: This is Excel's default chart type. The data is arranged horizontally with labels along the bottom of the chart (referred to as the **X axis**) and values vertically on the left hand side of the chart (referred to as the **Y axis**). The small box at the right hand side of the chart is the **legend** or **key** to the chart. When there is more than one set of data, a different colour will show which item is which on the chart.

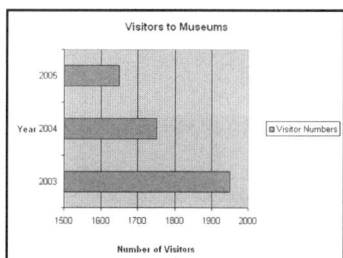

Bar Charts: This type of chart presents data in a similar format to the column chart, but in a sideways profile. The chart left shows the same data as the above column chart, but presents the data with the year labels on vertical axis and the number of visitor values on the bottom of the chart in increments.

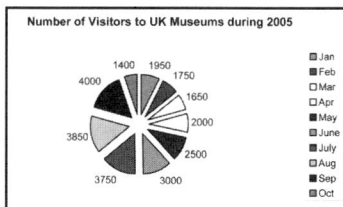

Pie Charts: These are mostly used to illustrate proportions of a whole, i.e. pieces of a pie. This example shows how the number of visitors' figures have been displayed in a different way to show the proportion of visitors per month over a year. Pie charts are only used to show one data series.

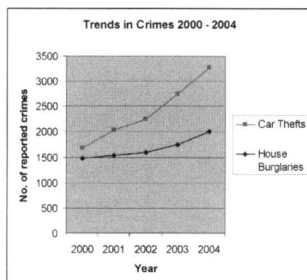

Line Charts: This type of chart is useful to show trends in data, e.g. rises or falls in sales figures, the example left shows trends in crime. There is usually more than one piece of information to the chart.

Changing a chart type

To change the chart type or style:

▶ Right-click anywhere within the chart and choose the **Chart Type** menu item from the pop-up menu.

▶ Select a different **Chart type** (or **Chart Sub-type** to change the style) from the **Chart Type** window.

> **Tip**
>
> You can preview how your data will look in a particular chart type by selecting a **Chart Type** option and clicking the **Press and Hold to View Sample** button.

Changing chart content

If you amend the content of worksheet cells on which your chart is based, Excel automatically updates the chart to reflect your changes.

Changing a chart's size

To resize a chart:

▶ Click once on the chart area.

▶ Click on any handle and hold down the mouse button as you drag the chart to a different shape.

> **Tip**
>
> If you drag on a corner handle, the chart expands and contracts proportionately to its current size; if you drag on an edge handle, the chart expands or contracts in that direction only. Excel automatically adjusts the font of chart text as you resize.

Changing a chart's appearance

To change the colour or format of any element in a chart:

▶ Click once on an element in the chart. For example, one of the columns on a column chart.

▶ Then **right-click** and choose the **format option** for the element. For example, on a column the choice would be **Format Data Point...**

Chart title

To edit the chart title:

▶ Click anywhere on the chart title to highlight the frame containing the title.

▶ Click anywhere within the title text. You can now edit the text.

To reformat the chart title:

▶ Double-click anywhere on the chart title to display the **Format Chart Title** window and select the options you require.

Data labels

To add a data label:

▶ Right-click on the **chart area**, e.g. a column or bar, and choose **Format Data Series** from the pop-up menu.

▶ Select the **Data Labels** tab. Here you can choose to show the numerical values or the names of the data points.

To reformat data labels:

▶ Right-click on a data label. This will select all of the data labels.

▶ Select **Format Data Labels** from the pop-up menu, and select your required options from the **Format Data Labels** window.

The example below, shows a chart with numerical data labels:

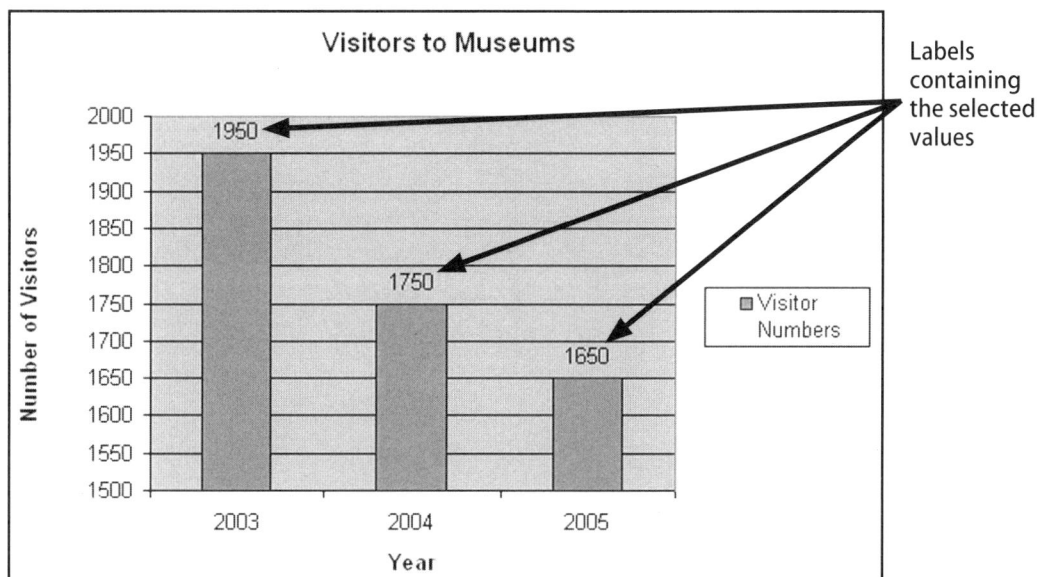

Labels containing the selected values

Chart scale

To change the scale of an axis:

▶ Right-click anywhere along the **axis** and choose **Format Axis...** from the pop-up menu.

▶ Select the **Scale** tab in the **Format Axis** window. Here you can change the minimum, maximum and increment values displayed for each axis, and the point at which the two axes cross.

Chart colours

You to change the colours of any part of a chart. In each case, you:

▶ Right-click on the element you want to change and choose the **Format** from the pop-up menu, e.g. **Format Axis...**

▶ If the element type selected is a **plot area**, **chart area**, or **chart element** (such as a bar), the **Format** window will display the **Patterns** tab. Here you can change the **Style**, **Color** and **Weight** using the options available on the drop-down menus.

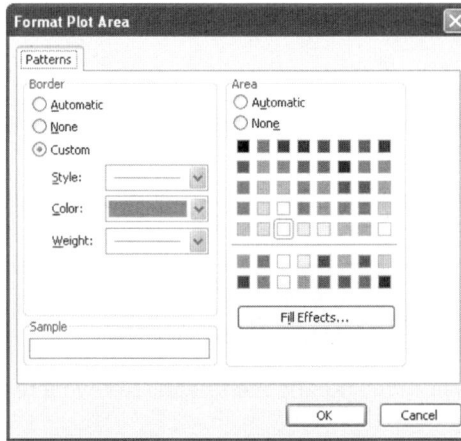

If the element type selected is a rule, e.g. an axis or grid, the **Format** window will display the **Patterns** tab. Here you can change the **Style**, **Color** and **Weight** using the options available on the drop-down menus.

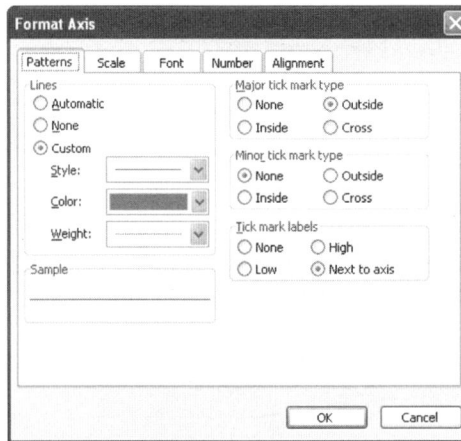

If the element type selected is text, e.g. a title or the labels on an axis the **Format** window select the **Font** tab. Here you can change the colour using the **Color** drop-down menu.

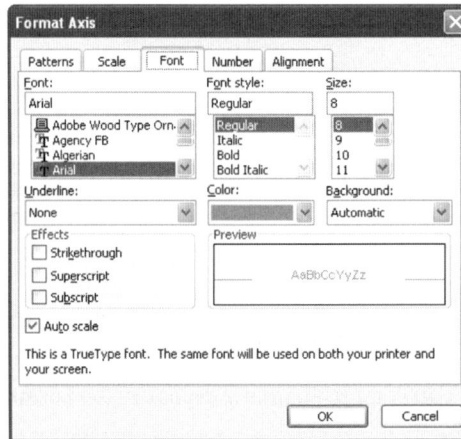

What is the best type of chart/graph to use?

The choice of chart/graph depends upon the type of information you are trying to display. If you want to display values as a proportion of a whole, e.g. the percentage of sales for different departments, then a pie chart would be useful to show the percentage share of the total sales that each department holds.

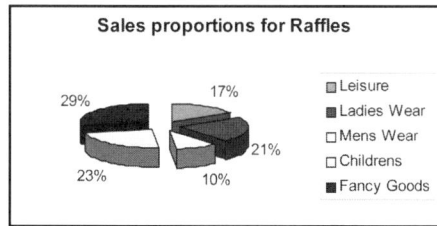

If you wish to compare values, e.g. one country's number of visitors against another, then a column chart would be suitable.

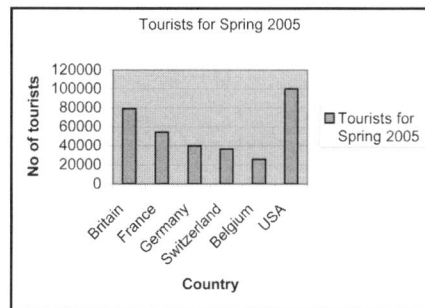

If you want to display trends of sales, etc., then you could use a line chart...

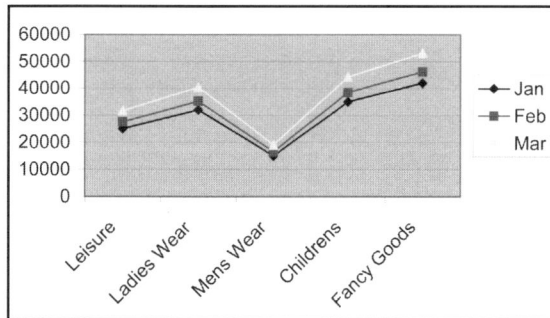

...or comparative bar chart.

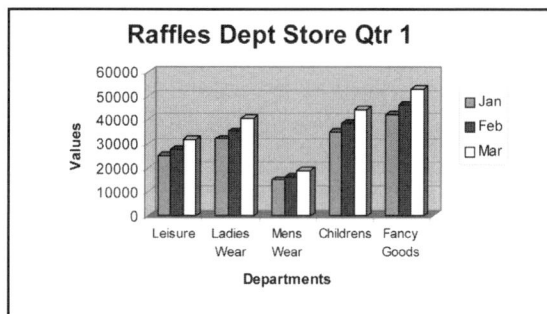

PRACTICE EXERCISE 10 – HIDING COLUMNS AND ROWS AND CREATING CHARTS

In this exercise we will make a **Column chart** that will show details of the first six coded items and their related **No. sold**.

1 Open the **Feathered Friends 3** spreadsheet you saved in **Practice Exercise 9**.

2 Hide columns **B – E** that appear between the two columns involved in the exercise.

3 Highlight the cells to be included in the chart (**A2 – F8**).

4 Create a suitable layout for the chart – experiment until you are confident all relevant data is included, the chart and its axes are correctly labelled and has an appropriate title.

5 **Print** the chart by highlighting the chart so that the black handles appear and select the **File** menu and then the **Print...** menu item.

6 **Highlight** the chart and **delete** it.

7 **Close** the spreadsheet application and **exit** from Excel.

DATABASES

What is a database?

A database is a collection of related information that can be organised in a structured way. You have probably used paper databases to look up information, e.g. a telephone book or an office filing system. If you are a music fan and have an extensive collection of CDs, MP3s or even LPs, you might want to create a database that contains information on your collection and each entry could include the title of the album, the artist, the year of release, the names of the individual tracks, the length of the individual tracks and the format. When you have collected all your data you will be able to keep track of your collection.

You will be able to sort the data so that you can see a list of all the artists in alphabetical order, or in order of the year of release. You could create queries that just show all the music you have on CD only and not display the information for music on the LPs or MP3s.

There are many uses for a database:

✓ The DVLA has a huge database of car owners, drivers and information relating to vehicles including car tax records.

✓ A library will also have a database in some form of all their books and who has borrowed them.

Both these examples relate to very complex areas. For the purpose of these tasks you will only need to set up one table with information (known as a **flat-file** database), which can be as simple as the names and addresses of the people in your class.

Parts of a database

✓ **Tables:** Information is stored in **tables**. Tables are made up of **fields**.

✓ **Fields:** These hold specific types of data such as a **date**, **text** or a **number**.

✓ **Records:** A record is made up of one or more pieces of data related to the same item. For example, the name, address, age, and date of birth could make up a record for a person.

✓ **Queries:** These are used to filter (or find) specific information contained in a database. For example, you could chose to view just the names of people and their date of birth who live in your town only.

The result of the query is displayed as a separate table view and can be saved with its own name and can be run at a later time. For example, when more records have been added.

Database structure

Before you can begin to enter information into a table you must design the structure of your database. You need to think about the names for the fields and the type of data that will be stored in each field, e.g. text, number, date, and you will need to consider the format of that data, e.g. if it is a number, how many decimal places should be shown? If it is a date, should it be a **short date** (19/12/06) or a **long date** (19 December 2006)?

First Name	Family Name	Address1	Address2	Address3	Post Code	Telephone Numb	Date of Birth	Salary
Marguerite	Patten	The Mill	Merseyside	Liverpool	L3 7AW	01254 668899	01/12/1954	£12,000.00
								£0.00

Employee Records : Table

Record: |◄| |◄| 2 |►| |►I| |►*| of 2

Database creation

▶ Launch **Microsoft Access**.

▶ Click on the **New** button.

▶ A **Task Pane** will appear on the right-hand side of the screen. Click on the **Blank database...** button.

🗐 Blank database...

▶ Select the location you want to save your database into.

▶ Access automatically provides a filename of **db1.mdb**, (or **db2.mdb**, etc., depending how many databases have been started). Give your database a useful name that you will remember. In this case we will call it **Personnel**.

▶ Click on the **Create** button.

1 Click on the new button.

2 Click on the **Blank database...** button.

3 Select the location to create your database in.

4 Choose a suitable filename.

5 Click on Create.

The Task Pane

▶ The **Database window** will now be displayed.

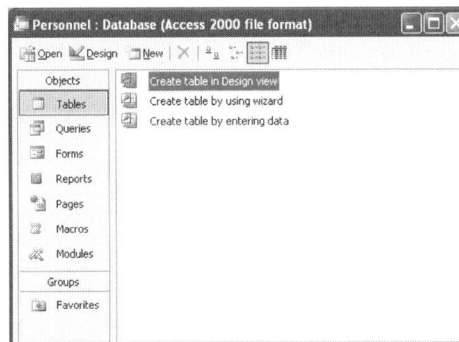

You are now going to create a **table**. You will add **field names** and **data types** for each field.

▶ Ensure that the **Tables object** is selected in the **Database window**.

> ☐ Tables

▶ Double-click on the **Create table in Design view** icon.

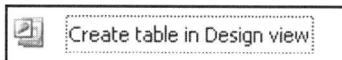

> Create table in Design view

▶ Use the following table to add the **Field Name** and **Data Type** for each field:

Field Name	Data Type
First Name	Text
Family Name	Text
Address1	Text
Address2	Text
Address3	Text
Post Code	Text
Telephone Number	Text
Date of Birth	Date/Time
Salary	Currency

> ### Note
>
> The **Address** field has the **Data Type** of text even though you may need to include house numbers and post codes. **Telephone Number** is also a text field as it has a space between the area code and the telephone number.

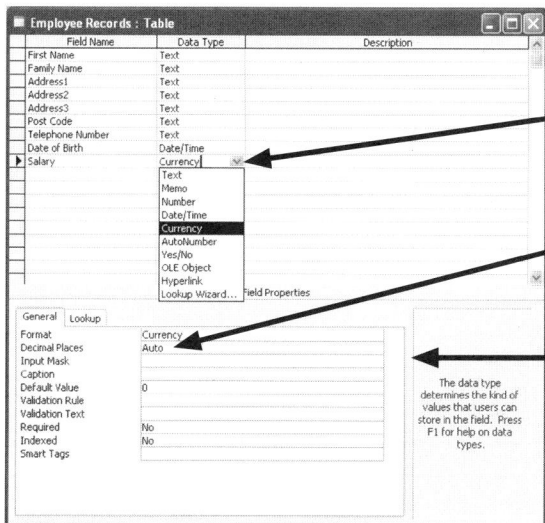

Select the **Data Type** from the drop-down menu.

Auto will display numbers to 2 decimal places.

The **Field Properties** box. Different options are displayed depending on the **Data Type** selected.

> ### Note
>
> In **Design View** the database fields are entered underneath each other, but when the table is opened in **Datasheet view** the fields will be displayed in columns across the screen.

▶ Click the **Save** button and give the table the filename **Employee Records**.

The following dialog window is displayed:

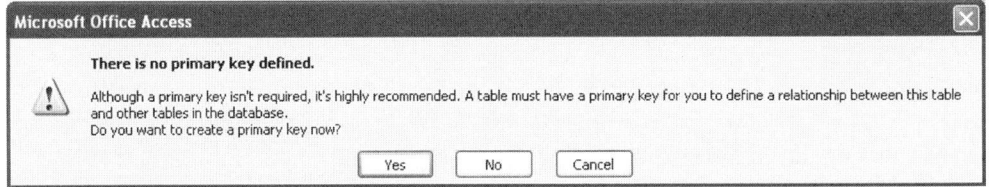

> A **Primary Key** is not needed for this example. Click **No**.

To ensure that each record in a table is unique, fields can be added that can contain a unique code or number to identify it. This is called the **Primary Key**. When you create a new database and you allow **Access** to insert a **Primary Key** field, rather than choose your own, each record is uniquely numbered in sequence beginning with number 1.

We are now ready to enter records into our new database. To do this:

> Ensure that the **Tables object** is selected in the **Database window**.

> Select the **Employee Records** table.

> Click **Open** to view the table in **Datasheet view**.

> Add the following data to the first record:

First Name	Family Name	Address 1	Address 2	Address 3	Post Code	Telephone Number	Date of Birth	Salary
Marguerite	Patten	The Mill	Merseyside	Liverpool	L3 7AW	01254 668899	01/12/1954	12000

Your first record will look something like this:

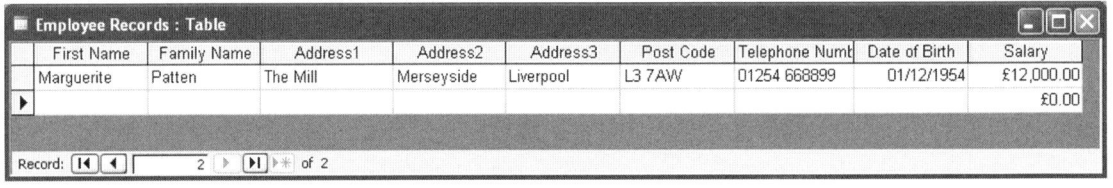

As you can see you will need to widen the table columns to be able to read your data.

> Position the mouse pointer between a column. You will notice the pointer change from an arrow to a cross with arrows pointing to the left and right.

Position the pointer between the columns ✛ and then double-click or drag the column to the size you require.

A Record is made up of fields.

Fields

Add six more records using the following data:

First Name	Family Name	Address 1	Address 2	Address 3	Post Code	Telephone Number	Date of Birth	Salary
Michael	Foster	23 The Grove	Brookside	Liverpool	L1 7ZZ	01254 447788	12/09/1933	23950
Jane	LeFevre	19 Orchard Road	Kilburn	Liverpool	L13 9QL	01254 665544	02/05/1985	9500
Darren	Eldrich	7 Rainhill Road	Whiston	Liverpool	L25 6VG	01254 882233	23/03/1985	10700
Conner	Oaks	25 Barley Road	Woolton	Liverpool	L11 4TT	01254 772255	29/02/1984	14200
Megan	Church	3 The Glade	Aintree	Liverpool	L31 3PP	01254 557010	24/12/1963	23500
Mark	Kirkup	18 Brooklands	Aintree	Liverpool	L31 3GP	01254 557407	22/09/1972	16750

Adjust the column widths again if necessary. Your table will look something like this:

	First Name	Family Name	Address 1	Address 2	Address 3	Post Code	Telephone Number	Date of Birth	Salary
	Marguerite	Patten	The Mill	Merseyside	Liverpool	L3 7AW	01254 668899	01/12/1954	£12,000.00
	Michael	Foster	23 The Grove	Brookside	Liverpool	L1 7ZZ	01254 447788	12/09/1933	£23,950.00
	Jane	LeFevre	19 Orchard Road	Kilburn	Liverpool	L13 9QL	01254 665544	02/05/1985	£9,500.00
	Darren	Eldrich	7 Rainhill Road	Whiston	Liverpool	L25 6VG	01254882233	23/03/1985	£10,700.00
	Conner	Oaks	25 Barley Road	Woolton	Liverpool	L11 4TT	01254772255	29/02/1984	£14,200.00
	Megan	Church	3 The Glade	Aintree	Liverpool	L31 3PP	01254557010	24/12/1963	£23,500.00
	Mark	Kirkup	18 Brooklands	Aintree	Liverpool	L31 3GP	01254557407	22/09/1972	£16,750.00
									£0.00

Employee Records : Table

Record: |◀ ◀ | 8 | ▶ ▶| ▶* | of 8

Click the **Close** button on the table window (not the Access window).

Click **Yes** if you are asked to save changes to the layout.

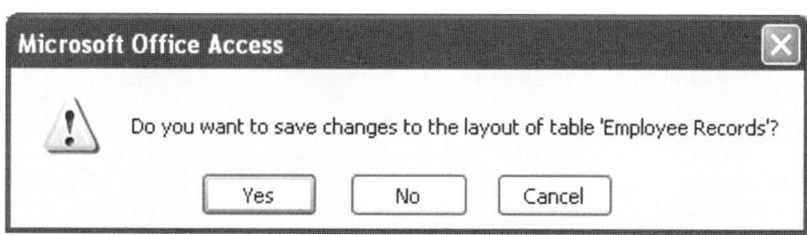

Microsoft Office Access

Do you want to save changes to the layout of table 'Employee Records'?

[Yes] [No] [Cancel]

Browsing records

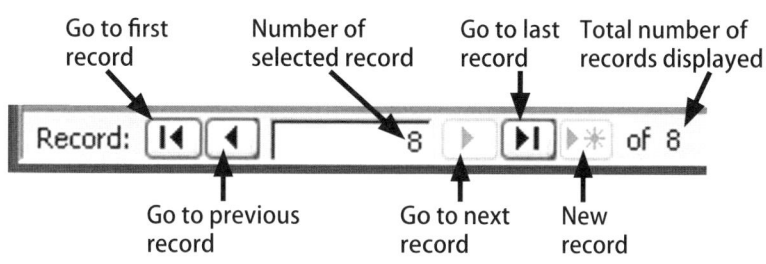

Go to first record | Number of selected record | Go to last record | Total number of records displayed

Record: |◀ ◀ | 8 | ▶ ▶| ▶* | of 8

Go to previous record | Go to next record | New record

Sorting records

It is useful to be able to sort your data on one particular field, for example by name, salary, date of birth, etc.

To carry out a simple sort:

➤ Open the **Personnel** database.

➤ Ensure that the **Tables object** is selected in the **Database window**.

| 🔲 Tables |

➤ Select the **Employee Records** table.

➤ Click **Open** to view the table in **Datasheet view**.

| 📇 Open |

➤ Select the **Salary** column by clicking on its heading.

lame	Address1	Address2	Address3	Post Code	Telephone Number	Date of Birth	Salary
	19 Orchard Road	Killburn	Liverpool	L13 90L	01254 665544	02/05/1985	£9,500.00
	The Mill	Merseyside	Liverpool	L3 7AW	01254 668899	01/12/1954	£12,000.00
	23 The Grove	Brookside	Liverpool	L1 7ZZ	01254 447788	12/09/1933	£25,950.00
							£0.00

▶ ▶I ▶✳ of 3

To select a column, click on the **column heading** containing the **field name**.

➤ Click on the **Sort Ascending** button on the **Table datasheet** toolbar and then the **Sort Descending** button and look at the results.

Sort Ascending button → [A↓ Z↓] ← Sort Descending button

> ### Note
>
> Using this type of sort it is only possible to sort on one field. For more complicated sorts you need to use queries. (*See Creating queries below.*)

➤ Click the **Close** button on the table window (not the Access window).

| ☒ |

➤ Click **No** when asked if you want to save the changes to the design of the **Employee Records**.

Creating queries

Queries are used to filter (or find) specific information contained in a database. In this example we will find employees with a salary of more than £10,000 and sort the data by the person earning the most.

➤ Open the **Personnel** database if you have not left it open.

➤ Select the **Tables object** in the **Database window**.

| 🔲 Queries |

➤ Double-click the **Create query in Design view** button.

| 📄 Create query in Design view |

⊘   The **Show Table** window is displayed. Select the **Employee Records** table and click the **Add** button.

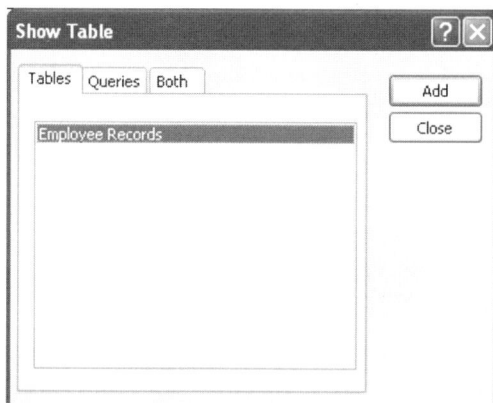

⊘   Click the **Close** button on the **Show Table** window.

The **Query design** window will be displayed ready to add fields:

Double-click on each field name in turn to add them to the query.

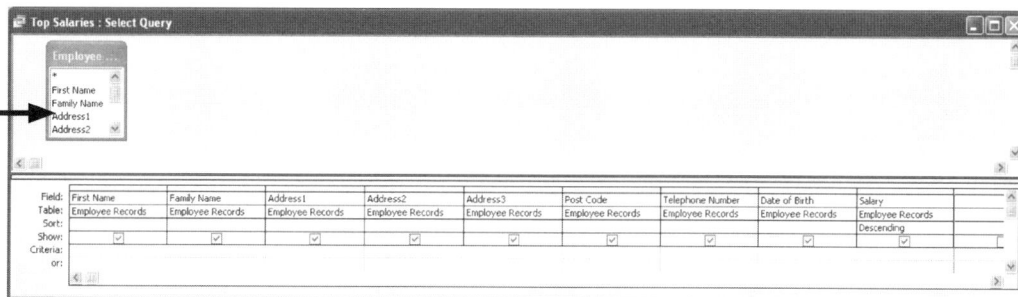

⊘   Double-click on each field name to add them to the query.

⊘   In the **Salary** field choose **Descending** from the drop-down menu in the **Sort** row.

⊘   In the **Criteria** row type a greater than sign > and **10000** (i.e. >**10000**).

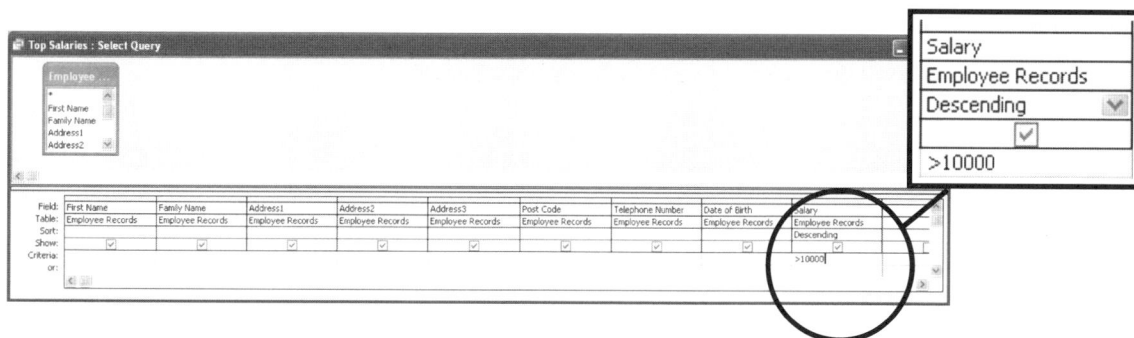

⊘   To run the query, click the **Run** button on the **Query Design** toolbar.

The results will appear as follows with the employee earning the most money at the top of the datasheet (anyone earning less than £10,000 does not appear in the query result):

First Name	Family Name	Address 1	Address 2	Address 3	Post Code	Telephone Number	Date of Birth	Salary
Michael	Foster	23 The Grove	Brookside	Liverpool	L1 7ZZ	01254 447788	12/09/1933	£23,950.00
Megan	Church	3 The Glade	Aintree	Liverpool	L31 3PP	01254557010	24/12/1963	£23,500.00
Mark	Kirkup	18 Brooklands	Aintree	Liverpool	L31 3GP	01254557407	22/09/1972	£16,750.00
Conner	Oaks	25 Barley Road	Woolton	Liverpool	L11 4TT	01254772255	29/02/1984	£14,200.00
Marguerite	Patten	The Mill	Merseyside	Liverpool	L3 7AW	01254 668899	01/12/1954	£12,000.00
Darren	Eldrich	7 Rainhill Road	Whiston	Liverpool	L25 6VG	01254882233	23/03/1985	£10,700.00
								£0.00

Record: 1 of 6

- Click the **Save** button.

- Name the query as **Top Salaries** in the **Save As** window.

- Click **OK**.

- Click the **Close** button on the **Query** window.

- Choose the **File** menu and then the **Exit** menu item to close Access.

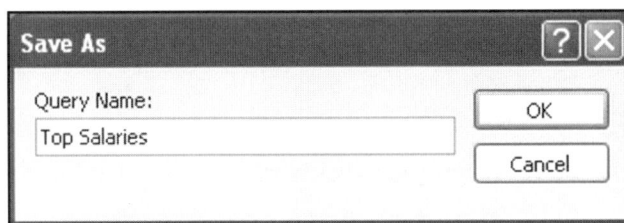

Save As

Query Name:

Top Salaries

OK   Cancel

Queries using multiple criteria

It is possible to query a database for more than one criterion at a time. In this example we will find the records of **First Names** starting with the letter **M** and earning **Greater than £20,000**.

- Open the **Top Salaries** query in **Design View**.

- In the **Criteria** for the **First Name** field enter **Like M*** (notice the use of the **asterisk** (*) **wildcard**. This is very similar to the wildcards used for finding files – *see page 143*.)

- In the **Criteria** for the **Salary** field change to **>20000**.

- **Run** the query

You should have two records displayed in the result (**Foster** and **Church**).

First Name	Family Name	Address1	Address2	Address3	Post Code	Telephone Number	Date of Birth	Salary
Michael	Foster	23 The Grove	Brookside	Liverpool	L1 7ZZ	01254 447788	12/09/1933	£23,950.00
Megan	Church	3 The Glade	Aintree	Liverpool	L31 3PP	01254 557010	24/12/1963	£23,500.00

Showing and hiding fields in a query

On some occasions you will find that you do not want to display a field in query results. For instance, you may want to print the results of the **Top Salaries** query, but you may need to hide the **Date of Birth** field. Rather than create another query you can simply hide the field.

- To **hide** a field select the **Design view** of the query, then click on the tick to **uncheck** the box.

- To **show** a field select the **Design view** of the query, then click on the empty box to **check** the box (add a tick).

Click here to remove the tick and hide the field. Click again to show the tick and redisplay the field.

Date of Birth
Employee Records
☑

Amending records

Sometimes the information in a record may need to be amended. For instance, a change of address, telephone number, salary or job title.

▶ Open the database containing the record(s) to be altered.

▶ Locate the field within record to be amended.

▶ Change the data and close the table.

> **Note**
>
> Access saves records without asking when you move to another record or close the table.

Inserting and deleting records

▶ Open the **Personnel** database.

▶ Ensure that the **Tables object** is selected in the **Database window**.

□ Tables

▶ Select the **Employee Records** table.

▶ Click **Open** to view the table in **Datasheet view**.

Open

▶ Select the record for **Michael Foster** by clicking on the grey box at the side of his name. The cursor will change to an arrow pointing to the right to indicate that the record can be selected. ➡

	First Name	Family Name	Address1	Address2	Address3	Post Code	Telephone Number	Date of Birth
	Marguerite	Patten	The Mill	Merseyside	Liverpool	L3 7AW	01254 668899	01/12/1954
▶	Michael	Foster	23 The Grove	Brookside	Liverpool	L1 7ZZ	01254 447788	12/09/1933
	Jane	Le Ferve	19 Orchard Road	Killburn	Liverpool	L13 90L	01254 665544	02/05/1985
*								

Record: I◀ ◀ 2 ▶ ▶I ▶* of 3

To select a record, place the mouse pointer over the **Record selector**. When the mouse pointer turns to an arrow, click. ➡

> **Tip:** If you need to delete more than one record hold the **Shift Key** down and click the records to be deleted.

▶ Press the **Delete** key.

The prompt screen below will appear for you to check that deletion is correct:

Microsoft Office Access

⚠ You are about to delete 1 record(s).

If you click Yes, you won't be able to undo this Delete operation.
Are you sure you want to delete these records?

[Yes]   [No]

▶ Click **Yes** and the table will now appear as follows:

	First Name	Family Name	Address1	Address2	Address3	Post Code	Telephone Number	Date of Birth
	Marguerite	Patten	The Mill	Merseyside	Liverpool	L3 7AW	01254 668899	01/12/1954
▶	Jane	Le Ferve	19 Orchard Road	Killburn	Liverpool	L13 90L	01254 665544	02/05/1985
*								

Record: I◀ ◀ 2 ▶ ▶I ▶* of 2

209

PRACTICE EXERCISE 11– CREATING AND SORTING A DATABASE

A children's nursery needs to keep records of the children with special dietary requirements.

1 Open Access and create a new database giving it a suitable filename.

2 Design a table using the data below. Use the column headers as the **Field Names**. All the fields, except **Age**, are to have the **Data Type** set as **Text** and set **Age** as **Number** with an **Integer** value.

Name of Child	Age (months)	Gender	Dietary Requirements	Attendance
Devlin, C	24	Male	Dairy Free	All Day
Sutton, M	11	Female	Gluten Free	All Day
Squires, D	21	Male	Dairy Free	All Day
Fletcher, D	8	Male	Gluten Free	Morning
Singh, K	9	Female	Vegetarian	Morning
Trent, L	17	Female	Dairy Free	All Day
Breeze, G	20	Female	Vegetarian	All Day
Sykes, V	18	Female	Dairy Free	All Day
Forbes, A	7	Female	Gluten Free	Morning
Westerbrook, J	30	Male	Vegetarian	All Day
Holt, A	29	Male	Gluten Free	All Day
Pincher, N	13	Male	Dairy Free	Morning

3 Enter the data as shown above.

4 **Sort** the records by **Dietary Requirements** in alphabetical order **A – Z**.

5 **Close** the database.

PRACTICE EXERCISE 12 – CREATING A QUERY

Ladybird Nursery wants to see how many children of one year and over have special dietary requirements and what those requirements are.

1 Open the database you created in **Practice Exercise 11**.

2 Create a query that includes the fields: **Name of Child**; **Age (Months)**; and **Special Dietary Requirements**.

3 Set the query to find all the records that have the **Age (Months)** with a value that is greater than months 11 (i.e. **>11**).

4 **Run** the query.

5 Check that the query results are as you would expect.

6 **Save** the query with a suitable name.

7 **Close** the database and **Exit** from Access.

Reports

Reports are used to display data, from a table or query, in a format that is easier to read and therefore more acceptable in business.

➤ Open the database you created in **Practice Exercise 11**. If you have not yet created it, you may need to in order to follow this example.

➤ Select **Reports** from the **Objects** list.

➤ Double-click on **Create report by using wizard**.

The **Report Wizard** window is then displayed:

Select the table or query you want to create a report for here. →

The available fields are displayed here. →

This button moves one field at a time to the **Selected Fields** box.

This button moves all fields to the **Selected Fields** box.

➤ We will use all the available fields in the report. The quickest way to add these is to click on the selector button that has two arrows. `>>`

➤ Click on the **Next** button.

It is possible to group records based on the content of one or more fields. We will group the data on two fields.

➤ Add the **Dietary Requirements** field to the report by selecting it and clicking on the right-arrow button or by double-clicking on the field name.

➤ Add the **Attendance** field to the form. Your screen should look like this:

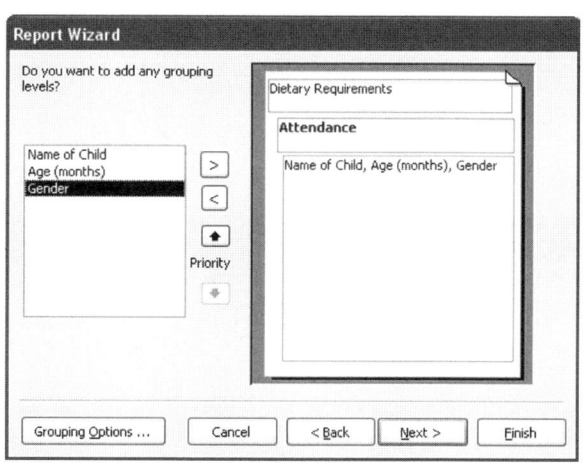

❯ Click on the **Next** button.

The next screen allows you to sort the data in **ascending** or **descending** order on one or more fields (depending on the number of fields you have in your database) if you wish. We will sort on two fields.

❯ Select **Name of Child** from the first drop-down list.

❯ Select **Gender** from the second.

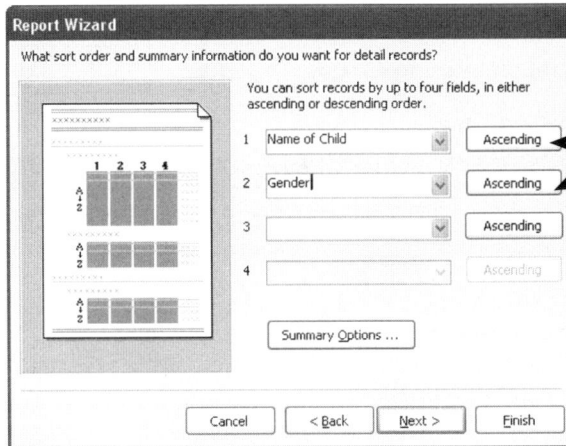

Click these buttons to change the sort order.

❯ Click on the **Next** button.

At the next screen you can choose the basic layout of the form.

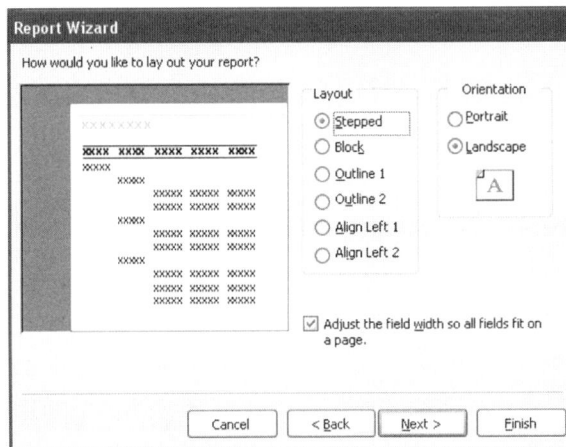

❯ Select **Stepped** as the **Layout** option.

❯ Select **Landscape** as the **Orientation** option.

❯ Ensure that the **Adjust the field width so all fields fit on a page** option is selected.

❯ Click on the **Next** button.

On the next screen you can choose a look for your report from a selection of designs.

▶ Select **Casual** from the list.

▶ Click on the **Next** button.

▶ On the final screen you can assign a name for the report. Leave this as the default name displayed or change it if you wish.

▶ Click on the **Finish** button.

▶ Check the preview of how the final report will look. Be careful to make sure that all the required information is displayed as you would expect.

In this example the word **Age** has not fitted on the page.

Special Dietary Requirements

Dietary Requirements	Attendance	Name of Child	Gender	(months)
Dairy Free				
	All Day			
		Devlin, C	Male	24
		Squires, D	Male	21
		Sykes, V	Female	18
		Trent, L	Female	17
	Morning			
		Pincher, N	Male	13
Gluten Free				
	All Day			
		Holt, A	Male	29
		Sutton, M	Female	11
	Morning			
		Fletcher, D	Male	8
		Forbes, A	Female	7
Vegetarian				
	All Day			
		Breeze, G	Female	20
		Westerbrook, J	Male	30

21 May 2006   Page 1 of 2

▶ Once you have noted any changes that you may need to make, click on the **Close** button.

Close

The new form will then be displayed in **Design view**, where you can fine tune the layout.

In **Design View** you will see that the form is divided into a number of sections:

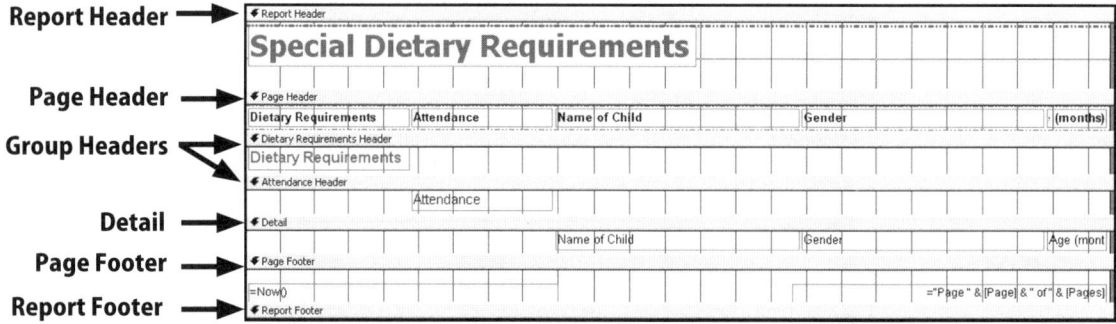

Report Header ➝

Special Dietary Requirements

Page Header ➝

Group Headers ➝

Detail ➝

Page Footer ➝

Report Footer ➝

Report Header – This appears on the first page of the report and contains the **Report Title**.

Page Header – This appears on every page of the report.

Group Headers – These are the fields you selected to group the data by.

Detail – This is where the remaining fields are displayed.

Page Footer – Appears at the bottom of every page.

Report Footer – Appears at the bottom of the last page of the report.

In order to read the word **Age**, in our example, we need to adjust the field labels in the **Page Header** for **Gender** and **Age (Months)**. To do this you will need to make the **Gender** label narrower and widen the **Age (Months)** label as follows:

- Select the **Gender** label so that there are eight black handles visible.

- Hover the cursor over the middle **left** handle until the cursor changes to a horizontal double-ended arrow. ↔

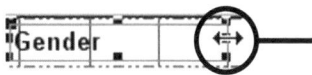

  Hover the cursor over the middle handle until it looks like this.

  > **Tip:** Right-click on a text label or field and select **Size** from the pop-up menu and then **To Fit** form the sub-menu.

- Click and drag the handle to the left to make enough room for the word **Age**.

- Select the **Age (Months)** label so that there are eight black handles visible.

- Hover the cursor over the middle **right** handle until the cursor changes to a horizontal double-ended arrow. ↔

- Click and drag the handle to the left until you think that there is enough room for the **Age** to be visible. Release the mouse button to check if **Age** has been displayed. If it hasn't, repeat the process again until it does.

- If you need to change font sizes of headings select the text label so that the eight black handles are visible and change the font size on the **Formatting Tool Bar**.

- When you are happy with the design click on the **Save** button.

- View the report by clicking on the **View** button.

- Check the layout again and repeat the process if necessary. Try to align headings and their text columns consistently.

INFORMATION AND COMMUNICATION TECHNOLOGY: PART A, PRACTICE TASKS

TASK DESCRIPTION GRID

Number and title	Page	Activities	Refer to reference sheet(s) on page(s)
1 Flappers	217	Editing a document.	156, 158, 159, 162, 170
		Producing an article.	
2 Costa Holidays	218	Editing a document.	156, 159, 157, 170, 162, 164, 171
		Producing a leaflet.	
3 Tiny Tots Nursery	219	Creating a table.	162, 163, 165, 156
4 Sunnydene Leisure	220	Conducting research.	146 – 150, 170, 162, 164
		Creating an illustrated calendar page.	
5 Bob Fixit	222	Creating an invoice in Excel.	177, 182, 184, 185 193, 183, 180, 181, 187, 188, 178,
6 Computer Knights	224	Creating a spreadsheet for costs and profits.	177, 181, 182, 184, 185, 183, 187, 188, 178
7 Aurora's Pantry	225	Creating a spreadsheet and graph.	177, 181, 183, 184, 185, 188, 187, 178, 193 – 200
		Taking a screen shot of the files created.	145
8 Costa Holidays	227	Producing a spreadsheet.	177, 183, 184, 187, 185, 178, 188
		Creating charts/graphs.	193 – 200
9 Rat Pack	228	Creating a database.	201 – 204
		Sorting and querying a database.	206 – 208
10 Wessex Windows	230	Creating a database .	201 – 204, 209, 206, 208
11 Daley Motors	231	Creating a database and a report.	201 – 204, 209, 206, 208, 211 – 214
12 Boneheart Surgery	233	Creating a database.	201 – 204
		Sorting and querying a database.	206 – 208
		Producing a report.	211 – 214
13 Uwin Solicitors	234	Internet research.	146 – 150
		Producing a Word document.	
		Sending an email with a file attachment.	153 – 154

Number and title	Page	Activities	Refer to reference sheet(s) on page(s)
14 Feel Good Factor	235	Internet research.	146 – 150
		Producing a Word document.	
		Sending and receiving emails with file attachments.	153 – 154
15 Fitness	236	Creating a data source file.	172 – 176
		Carrying out mail merge.	
16 Green Fingers	239	Creating and completing a spreadsheet.	177, 179, 181, 182, 183, 184, 185, 187, 188, 193, 178, 162, 164, 163
		Creating an illustrated letter heading.	
		Creating a diary sheet.	
		Writing a statement related to dealing with hardware and software problems.	
17 Just Minute	242	Creating, sorting and querying a database.	201 – 208
		Producing a database report.	211 – 214
		Editing a spreadsheet	181, 182, 183, 184, 185, 187, 188, 193
		Carrying out a mail merge.	172 – 176
		Taking a screen shot of the files created.	145

TASK 1: FLAPPERS

Student Information	REMEMBER:
In this task, you will prepare an article on the history of fashion using appropriate formatting techniques. Ask your tutor for a copy of the document **1920s Fashion.doc**.	To save your document regularly. To add your name and the date to the footer before printing. *Refer to pages 156, 158, 159, 162, 170* for help with this task.

Editing a document

Scenario

You are involved in recalling a document prepared by someone else and editing it in order to prepare a finished document with an image.

Activities

1   Open the Word document called **1920s Fashion.doc**.

2   Select the entire document and change the font to **Arial**.

3   Make sure the page orientation is **portrait**.

4   Change the **font size** of the text **Fashion in the 1920s** to **18 point** and change it to **All Caps**.

5   **Align** the main heading so it is in line with the left margin for the whole document.

6   **Move** the paragraph beginning "**In 1924.......**" so that it appears at the end of the document.

7   Add your name, **right-aligned**, at the end of the page after the last paragraph.

8   Display **bullets** at the beginning of the six main paragraphs.

9   Change the line spacing of the whole document to **1.5 line spacing**.

10 **Justify** the six paragraphs.

11 Insert a **table** at the bottom of the document below your name – **1 row by 2 columns**.

12 Enter the following text:

Date Received:	Date Checked:

13 Insert a **page number**, centred in the **footer** of your document.

14 Insert an appropriate image at the bottom of the page below the table. Place this image on the right-hand side of the page.

15 **Resize** the image to **2.5cm x 3.5cm** and move it so it ends in line with the right margin.

16 Use the **Spelling and Grammar** checker to make necessary changes.

17 **Save** the document with the filename **Fashion** and print **one** copy.

TASK 2: COSTA HOLIDAYS

Student Information	REMEMBER:
In this task, you will edit an information leaflet about Venice. Ask you tutor for a copy of the **Venice.doc** file and the **rialto.jpg** image file.	To save your document regularly. To add your name and the date to the footer before printing. *Refer to pages 156, 159, 170, 162, 164, 171 and 157* for help with this task.

Editing a document

Scenario

You work for a travel agent, **Costa Holidays**, and you have been asked by a colleague to have a look at the leaflet he has drafted and he has asked you to make amendments and improve its layout.

Activities

1   Open the document called **Venice.doc**.

2   Select the entire document and change the **font** to **Arial**.

3   Make the heading, **Venice**, **bold**.

4   **Left align** each of the three subheadings in the document.

5   Change the subheadings to **uppercase**.

6   Change the **font** of the heading to **Arial Black, size 12**.

7   Add **bullets** to each of the subheadings – **Transport**, **St Mark's Square** and **Landmarks**.

8   Change the **line spacing** of the whole document to **1.5**.

9   Working on the first paragraph, insert spaces before each of the following so that separate paragraphs are formed:
Just like anywhere ....
Strictly ....

10   Insert the **rialto.jpg** image at the top of the page, placed centrally under the heading **Venice**.

11   Use the **Spelling and Grammar** checker to make necessary corrections and check thoroughly yourself.

12   Insert a table, two columns by three rows, at the end of the document containing the following text:

Transport	Water buses
Landmarks	Ponte Rialto
Attractions	St Mark's Basilica

13   Modify the table **borders** so that the lines are **2¼pt**.

14   If necessary, change the size of the image and/or font size to make sure the article appears on **one page only**.

15   Insert a **footer** that contains the following information fields:

Your name   *The new document filename*   *Today's date*

16   **Save** the document, using the filename and **print** a copy.

TASK 3: TINY TOTS

Student Information	REMEMBER:
In this task, you will produce an information leaflet.	You will need to maintain a consistent format. You will need to check that all font styles and sizes are suitable. *Refer to pages 162, 163, 165 and 156 for help with this task.*

Producing a table in Word

Scenario

You work in a children's nursery called **Tiny Tots** and today you have to prepare a document that will help staff to monitor children's well-being whilst at the nursery.

Activities

1   Create a new Word file and create the following table, formatting it and the text as shown:

<u>Active</u> The amount of physical activity on various tasks and in various settings.	
Low Activity	High Activity
Play is physically calm, sits still during meals and while watching television. Bath time is quiet.	Physically active in play, active during mealtimes, while watching television, during bathing and at bedtime.
<u>Regular</u> Regularity of biological functions	
Irregular	Regular
Falls asleep at different times. Difficult to know how long will stay asleep. Hungry at different times and eats varying amounts of food. Unpredictable	Goes to sleep within an hour of about the same time every night. Eats about the same amount of food each day. Is hungry at about the same time each day. Predictable
<u>Expresses Emotions</u> The intensity in emotional expression.	
Low Intensity	High Intensity
When unhappy looks downcast, may whimper or whine or cry quietly. When happy, smiles, chuckles or giggles. When angry, looks cross or talks a little louder than usual.	Cries loudly when unhappy. Roars with laughter, runs around excitedly, shouts with joy. When angry, may scream shout and jump about.

Continued

Enhancing the look of the table

2   **Merge** the first row containing the **Active** information so that the text is spread across the width of the table, i.e. there is **one** cell across the two columns.

3   Repeat this for **Regular** and **Expresses Emotions**.

4   Change the appearance of any words you feel should be **emphasised** in the text.

5   Use any combination of **shading**, **borders** and **font styles** to make the document look more attractive and easier to read.

6   Add a **footer** to the document that includes the following information fields:

Your name   *The new document filename*   *Today's date*

7   **Save** the document with a suitable filename and print a copy.

TASK 4: SUNNYDENE LEISURE

Student Information

In this task, you will be undertaking research on a sporting event and creating a calendar page. Ask your tutor for the **Sunnydene.bmp** image file.

REMEMBER:

You will need to obtain some research material from either the Internet and/or a reference book that will include text and images to support your work.

You should demonstrate that you can create and adapt structures for the development of text, number and image.

You should also take any opportunity to improve the layout by using formatting techniques available.

You will need to make up an address and contact details.

Refer to pages 146 – 150, 170 and 162 and 164 for help with this task.

Researching information, creating a document, writing a statement about copyright

Scenario

You work at **Sunnydene Leisure plc** and your task today is to put together a one-page document representing one month of next year. This page, when approved, will go into next year's calendar.

Activities

Drafting your calendar page to show the logo and the sporting information

1   Arrange the company's logo and contact details however you wish on the top half of the page (**Sunnydene. bmp** file). Following the logo and address will be some details of the sport you choose to represent the month, and an appropriate image that represents the sport.

2   Create a new Word document and insert the logo and contact details.

3   Conduct Internet-based research into main sporting events in the UK.

Note: You must match the sport to the month that will appear on the page. For instance, if you select test cricket you will illustrate the month of June or July, which is the test cricket season. If you select tennis, you will use the month of June. If you select the European Football Cup, then June is the relevant month. If you select horse jumping, the month might be October, etc.

4   **Copy** the text and image(s) that you find and want to adapt into Word.

> ### Note
>
> Some images on the Internet are subject to copyright so you may find that you cannot select them from the web.

5   **Print** a copy of the Word document(s) you create before you adapt them so your tutor can check the text in your calendar page has been adapted and not merely copied.

6   Adapt the text and combine the sporting image and, when you are happy with it, copy into your calendar page, underneath your company's logo.

7   Adjust the content and appearance as necessary, **remembering** it must only take up the top half of the page.

Inserting your calendar month

8   The final step is to add the calendar dates in the bottom half of the page. This can be done by adding a table.

            You will need seven columns and six rows.

            The table must have the days of the week in the first row, as shown below:

Monday	Tuesday	Wednesday	Thursday	Friday	Saturday	Sunday
		1	2	3	4	5
6	7	8				
				31		

9   Consult a diary or calendar for next year so that you have the **correct** days matched to dates for your sporting month.

10   Fill in the dates in your table. For instance, in the example the 1st of the month begins on a Wednesday and the 31st is a Friday. These dates have been added appropriately and the unoccupied days shaded.

11   Experiment with your layout until you are happy it is attractive and **easy to read**.

12   Review the layout of your calendar page, **save** the file and **print** when complete.

13   Because you have worked with images that have been copied from the Internet, write a statement of about 150 words, describing what copyright is, what the Law says, and why it is important you do not break the **Copyright Law**.

> ### Remember
>
> Hand in your research (source) documents with the task.

TASK 5: BOB FIXIT

Student Information	**REMEMBER:**
In this task, you will create an invoice template so details and costs can be added to produce invoices for the firm.	You will need to ensure formatting is consistent. You will need to think about the formulae you will use. *Refer to pages 177, 178, 180 – 185, 187,188 and 193* for help with this task.

Creating a spreadsheet

Scenario

Bob Fixit is a self-employed builder. He has been doing his paperwork by hand, but his wife has persuaded him to invest in a computer to make life easier. Today you are setting up a spreadsheet and using it to prepare an invoice.

Activities

1   Create a new blank workbook and design an invoice sheet that contains the information illustrated in **Appendix 1**, beginning in **cell A1**.

2   **Merge cells**, experiment with **outline** and **inside borders**, **extend row widths**, use **bold** and **font sizes** and **styles** to produce an attractive invoice form.

3   Add an appropriate image next to the title **Invoice**.

4   Add a custom footer that includes:

   Your name   *The new document filename*   *Today's date*

5   Return to the spreadsheet you saved, opening it if you closed it, and prepare to enter details of an invoice for a customer. (Details have been handwritten by Bob on a note pad – **Appendix 2**).

   The date is today's (dd/mm/yyyy).
   Invoice Number: BF176/5

6   In the shaded cells enter the appropriate formulae to calculate the **Total excluding VAT**, the **VAT** and the **Total Due**. Format the column as **Currency** showing the £ symbol and to **2 decimal** places.

   The table below gives an example of the formulae you need:

Total (excluding VAT)	=SUM(C10:C25)
VAT @17.5%	=(Total Exc VAT cell)*17.5%
Grand Total	=(Total + VAT) cells

   Note: The cell references you have created are likely to be different to the example shown above.

7   **Save** the Invoice with these calculations, using a different filename as you do not want the template created in Activities 1 – 4 to be overwritten.

8   **Print** two copies, one showing the **values** and one showing the **formulae**.

9   **Close** the file and exit from Excel.

Bob Fixit – Builder Unit 25, Frosterly Trading Park, Ross on Wye, Hereford HE5 8BT 01453 334457

INVOICE

| To: | | | Date: |
| | | | Invoice No: |

Date	Description of Work		Price
	Total (excluding VAT)		
	VAT @17.5%		
	Total Due		

Please send an Invoice to: Mr and Mrs J Pleasance 12 Linden Villas Ross on Wye Hereford HE5 9RR

Work done on date 11th May

Repairs to roof damaged in storm

Tiles   £65.00

Plaster and sealant   £35.60

Labour   £65.00

Bob (yesterday's date)

TASK 6: COMPUTER KNIGHTS

Student Information	REMEMBER:
In this task, you will create, and complete, a spreadsheet.	Read all instructions carefully. Check your answers are accurate. *To refer to pages 177, 178, 181 – 185, 187 and 188 for help with this task.*

Creating a spreadsheet

Scenario

You work for **Computer Knights** and in this task you will prepare a spreadsheet and add formulae so that the company can determine how much profit it will make on each item for sale.

Activities

1. Create new blank Excel workbook and enter the data for **columns B to G**. Format the columns and wrap the text appropriately.

	A	B	C	D	E	F	G
1	Part	Item Description	Cost	VAT	Purchase Price	Selling Price	Profit
2	Accessories	AB3 Wireless mouse	£20.00				
3		RK 200 cordless optical keyboard and mouse	£19.99				
4		Standard Keyboard	£4.99				
5		USB Portable Flash Drive	£25.45				
6		CD Rom 10 Pack	£4.95				
7		Floppy Disks 20 Pack	£10.99				
8	Printers	HP1223 Deskjet Super	£78.50				
9		AL999 Laserbeam Excelsior	£125.00				
10		EP655 Inkjet Basic	£45.95				
11		SP11 All in one Printer, Fax and copier	£75.99				

2. Add the word **Part** in **cell A1** and change the text direction as indicated, selecting **Horizontal Center** and **Vertical Center** for the text alignment options.

3. Add the word **Accessories** in **cell A2** and change the text direction and alignment, merging **cells A2 – A7 inclusive**.

4. Add the word **Printers** in **cell A8**, change the text direction and merge **cells A8 – A11** inclusive.

5. Add a formula that will calculate the **VAT**, which is charged on all goods at **17.5%** added to the price (**Cost** multiplied by **17.5%**), for the **AB3 Wireless mouse** then **copy/replicate** the formula to the remaining items..

Continued

6   Add a formula to calculate the **Purchase Price**, which will be the **Cost Price** plus the **VAT**. Remember to copy the formula to each item.

To calculate the **Selling Price** you must add **10%** to the **Purchase Price** (i.e. **Purchase Price multiplied by 10%**). Replicate it for all items.

7   The final calculation is **Profit**. Add a formula to calculate this, by deducting **Purchase Price** from the **Selling Price**. Replicate it for all items.

8   Develop the presentation of the spreadsheet by using shading, emboldening etc., to enhance entries. Remember to be consistent with the presentation of currency in each column.

9   Add a **custom footer** that will show the date, your name and the filename.

10 **Print** a copy of the spreadsheet **twice**, one to show the **values** and one to show the **formulae**.

11 **Close** Excel.

TASK 7: AURORA'S PANTRY

Student Information
In this task, you will prepare a spreadsheet that uses the **average** formula, and then produce a chart (graph) from selected information.

REMEMBER:
Read all instructions carefully.

Check your answers are accurate.

To refer to pages 177, 178, 181, 183 – 185, 187, 188 and 193 – 200 and 145 for help with this task.

Creating a spreadsheet and a chart

Scenario
Mr George Star, the owner of **Aurora's Pantry**, has given you details of the number of items sold between January and June this year. He has asked you to create a spreadsheet that will monitor the average takings over the **last** six months of trading.

Activities
1   Create a new blank Excel workbook and enter the information shown in **Appendix 1**, beginning in **Row 1** with the title **Aurora's Pantry**.

Use appropriate, borders, shading and formatting and be consistent in your formatting of figures and text. **You might need to make your text contained in the shaded cells a different colour**.

2   Enter formula that will calculate the **Total** of each item sold.

3   Enter formula that will calculate the **Average** number of sandwiches sold from January to June. Copy the formula for each item sold.

4   Add a formula that calculates the items that were the best sellers (**Max**) and that were the worst sellers (**Min**) in each month.

5   **Print** two copies, one showing the **formulae** and one showing the **values**.

6   Add a **custom footer** that shows:

Your name   *The new document filename*   *Today's date*

Continued

7   **Print** two copies, one showing the **formulae** and one showing the **values**. **Make sure all the information is visible**.

Creating a chart/graph

In January and April, toast proved to be a constant best seller whilst salads were the worst sellers over the same period.

8   Produce a suitable chat to show the trends of all items over the same period (January and April) using the **Chart wizard** in Excel.

9   **Label** the **axes**. Add the chart title **Sales trends for Jan and Apr**.

10   **Print** the chart.

11   **Save** the document appropriately.

12   **Close** Excel.

13   Take a screen shot and print the image of the directory showing the filenames you created in this task.

Appendix 1

Aurora's Pantry

Items	Jan	Feb	Mar	Apr	May	Jun	Total	Average
Sandwiches	42	35	36	45	50	60		
Scones	21	18	17	35	25	27		
Soup	52	60	45	52	22	20		
Salads	0	5	10	15	54	65		
Muffins	45	52	65	70	65	65		
Cakes	32	14	35	54	75	99		
Pastries	61	21	65	55	63	87		
Teacakes	75	37	61	55	31	68		
Toast	85	92	75	77	25	44		
Total								
Best Seller								
Worst Seller								

TASK 8: COSTA HOLIDAYS

Student Information	REMEMBER:
In this task, you will create a spreadsheet to calculate the average and total costs. In addition, you will develop the figures to display the costs in suitable charts.	By default text is aligned to the left and number to the right within a spreadsheet. All numerical data should have a consistent format. All cells are referred to by the column heading then row number, e.g. A1. *Refer to pages 177, 178, 183 – 185, 187, 188 and 193 – 200 for help with this task.*

Creating a spreadsheet and charts

Scenario

The manager of **Costa Holidays**, Abel da Costa, has given you some data on the main holiday destinations sold by the company. He would like you to calculate total and average costs. In addition, he needs charts (graphs) for a presentation he has to give to the Board of Directors.

Activities

1   Create a new blank Excel workbook and enter the data shown below, beginning at **cell A1**. Format **cells B3** to **E6** as **currency** showing the **£** symbol and **no** decimal places.

Costa Holidays				
Destination	Algarve	Benidorm	Cyprus	Average
Hotel	215	245	285	
Flight	154	200	245	
Insurance	35	30	40	
Total Cost				

   Note: The shaded cells will have formulae entered into them later in the task. You can shade them if you so wish.

2   **Embolden** the content of **cell A1** and merge this row across **cells A1 to E1**.

3   **Centre** each column heading.

4   Add a formula to **cell B6** that will total the **Algarve** column.

5   Copy the formula for **Benidorm** and **Cyprus**.

6   Ensure the formatting of these cells is consistent.

7   The next step is to calculate the **average price** for each of the costs across all the destinations.

    Enter the formula =**average(B3:D3)** in **cell E3**.

    Copy the formula down to the **Insurance** row.

8   Add a custom footer that shows:

    Your name   The new document filename   Today's date

9   **Print** two copies: one showing the **formulae**, and one showing the **values**.

Continued

Converting spreadsheet information into charts

10   The first chart will be a **column chart** to show the distribution of costs for all of the destinations. Select the cell range **A2:D5** and create a **column chart**, with the title **Distribution of Costs**.

11   **Print** the chart. You can decide whether or not to include the spreadsheet in the printout. After printing, delete the chart.

Next you will produce a **bar chart** that represents the costs of:

     The Algarve and Cyprus: Hotel, Flight and Insurance.
     You will need to **hide** the column containing the information on Benidorm.

12   Create the **bar chart** using the heading **Comparison of Holiday Costs to the Algarve and Cyprus**.

     Label the axes **Cost** and **Destination**.
     Add **Data Labels** for the **Series Name**.

13   **Print** the chart, or spreadsheet and chart.

14   **Close** Excel.

TASK 9: RAT PACK

Student Information	REMEMBER:
In this task, you will create a database.	You need to identify data types and fields before beginning your database. Your database table should be printed in landscape format. *Refer to pages 201 – 204 and 206 – 208 for help with this task.*

Creating, sorting and querying a database

Scenario

Today you work for **Rat Pack**, a small retailer of musical instruments. You have been asked to create a database to enable the business to monitor its stock levels.

Activities

1   Open Access and create a new database and name it **Rat Pack**.

2   Create a new table and add the **Field Names** and appropriate **Data Types** using the information in the following table:

Field Name	Data Type/Properties
Instrument	Text
Date in Stock	Date/Time, Short Date
No in Stock	Number
No Sold	Number
Price	Currency, £ sign, 2 decimal places
Re Order	Yes/No

Continued

3   **Save** the table with the name **Rat Pack Stock - Your Name** (e.g. Rat Pack Stock - Jack Gray).

4   Enter the information from the table below into the database:

Instrument	Date in Stock	No in Stock	No Sold	Price	Yes or No
Trombone	01/03/2006	10	5	£98.50	No
Tenor Saxophone	05/03/2006	2	1	£66.80	Yes
Clarinet	10/03/2006	6	2	£42.50	No
Bass Trombone	06/03/2006	12	6	£59.75	No
Alto Saxophone	30/03/2006	8	7	£42.00	Yes
Oboe	02/02/2006	10	4	£24.00	No
Drums	08/02/2006	12	5	£199.00	No
Classical Guitar	10/02/2006	3	1	£99.00	Yes
Cornet	14/02/2006	6	2	£32.45	No
Trumpet	27/02/2006	4	1	£50.00	Yes

5   **Sort** the table in order of **Date in Stock** with the **earliest** date appearing at the top of the list.

6   **Print** a copy of the sorted table.

7   Create a query to find all items that need **reordering** (i.e. returning **Yes**). Include **all** the fields in your query.

8   **Save** the query as **Reorder Stock - Your Name**

9   **Print** the query results.

10   Prepare another query that will show which instruments can be purchased for **£50 or less**. Include only the fields **Instrument: Price: No in Stock** in that order.

11   **Sort** the query into **descending** order of **price**.

12   **Save** the query as **Stock costing £50 or less – Your Name**

13   **Print** the results.

14   **Exit** from Access.

Note

If you have selected a **Data Type** of **Yes/No** for the **Yes or No** field, you will need to **check** the box so that a tick ☑ is displayed for **Yes**, otherwise leave the it blank for **No**.

Tip

Include your name when naming Tables, Queries and Reports in your database to make it easier to identify any documents when they are printed.

TASK 10: WESSEX WINDOWS

<table>
<tr><td>

Student Information

In this task, you will create a database, add, change, delete records and create a query using multiple criteria.

</td><td>

REMEMBER:

Identify data types and fields before beginning your database.
The database fields are entered vertically in design view, but change to column headings in table view.
Your database table should be printed as landscape format.
Refer to pages 201 – 204, 206, 208 and 209 for help with this task.

</td></tr>
</table>

Creating, amending, sorting and querying a database

Scenario

In this task, you will create a prototype database for the personnel department of **Wessex Windows** to demonstrate how the department could hold information about its employees.

Activities

1   Open Access and create a new database and name it **Wessex Personnel**

2   Create a new table with **Field Names** and appropriate **Data Types** for the information in the following table:

Surname	Initial(s)	Department	Gender	Grade
Green	F	Sales	M	1
Whyte	JG	Administration	M	1
Black	B	Sales	M	2
Browne	LD	Sales	F	2
Pinkerton	P	Warehouse	F	1
Redding	O	Warehouse	M	3
Gray	J	Marketing	F	3
Scott	KMB	Marketing	M	3
Walters	T	Administration	F	1
Bates	W	Warehouse	M	2
Pinter	W	Administration	M	2
Gibson	A	Sales	F	1
Lake	B	Administration	F	3
Carter	T	Sales	M	2
Benson	R	Warehouse	M	3

3   **Save** the table as **Wessex Staff - Your Name**. Open the table and enter the data as above.

4   **Print** the table.

5   Add the following records into the database as three new employees have joined the company:

Bell	K	Sales	F	1
Rathbone	D	Warehouse	M	3
Lewis	C	Admin	F	1

Continued

6   Find the record for **B Black** and change the **Grade** to **3**, as he has been promoted.

7   **Delete** the record for **J Gray**, as she has left the company.

8   **Sort** the records into **alphabetical order (A – Z)** on **Surname**.

9   **Query** the database, including all fields, to show all the people who work in the **Warehouse** whose **Grade** is more than **1**.

10   **Save** the query as **Senior Staff - Your Name**

11   Prepare to print the query results **not displaying** the **Gender** field.

12   **Print** the query results.

13   **Exit** from Access.

TASK 11: DALEY MOTORS

Student Information	**REMEMBER:**
In this task, you will create a database, create a query using multiple criteria and produce a report.	You need to identify data types and fields before beginning your database. The database fields are entered vertically in design view, but change to column headings in table view. Your database table should be printed in landscape format. *Refer to pages 201 – 204, 206, 208, 209 and 211 – 214 for help with this task.*

Creating a database and a report

Scenario

You work for the showroom manager of **Daley Motors** and today have to produce a database of cars the company has bought, or agreed to buy. You must then produce queries and a report.

Activities

1   Open Access and create a new database. Name the database file as **Daley Motors**

2   Create a new table named **Purchases - Your Name** and add **Field Names** with appropriate **Data Types** for the information in **Appendix 1**.

3   **Sort** the table into **alphabetical order (A – Z)** on **Model**. **Print** a copy of the table.

4   Create, and print, the following, **separate**, queries:

   a   Cars with mileage less than **50,000**. Name the query **Low Mileage - Your Name**

   b   Cars with only **1** previous owner, sorted in ascending order of **Year**. Name the query **One Owner - Your Name**

   c   Cars made after **1999** worth less than **£10,000** and not paid for. Sort into **descending** order of **Price**. Name the query **To Purchase - Your Name**

Continued

5   Create a **report** from the last query (**c**). Include the fields:

Price; **Reg No**; **Model**; **Make**; **Year**.
Sort into descending order of **Price**.
Select any type of layout design.
Use the **Report Title**:
Cars not paid for and manufactured after 1999 - Your Name
Include **today's date** in the **Report Footer** and **Page 1 of 1** (this is usually the default setting).
Check all information is visible, adjusting the layout if necessary.

6   **Print** the report in **landscape**.

7   **Close** the database.

Appendix 1

Purchases - Your Name									
Reg No	Model	CC	Year	Mileage	Colour	Make	Previous Owners	Price	Paid
379 CZ	QUATTRO	3500	1996	15082	MAROON	AUDI	2	£6,775.00	No
B123 FJK	250	2500	1999	57950	BLACK	VOLVO	3	£2,795.00	No
CUP 270W	WHIZZKID	970	1990	54322	BLACK	SUZUKI	6	£300.00	Yes
D325 SJR	BLUEBIRD	2000	1995	28960	BLUE	NISSAN	4	£3,895.00	No
BT52 ASX	TURBO	3500	2002	8764	BLACK	SAAB	1	£12,750.00	No
DE52 GWZ	200	1400	2002	21788	GREEN	ROVER	1	£8,677.00	Yes
E567 DAW	XJ6	6000	2003	19887	SILVER	JAGUAR	1	£11,999.00	No
LC53 TYB	LAND CRUISER	2000	2003	13564	BLUE	TOYOTA	1	£16,788.00	Yes
ELC 13	VECTRA	2000	1999	108465	GREEN	VAUXHALL	2	£8,900.00	Yes
Y463 JJM	A CLASS	1600	2000	26934	WHITE	MERCEDES	1	£7,699.00	No
NU04 YTB	C3	1400	2004	8763	BLACK	CITROEN	1	£7,800.00	Yes
EL03 LGL	YARIS	1400	2003	14670	GOLD	TOYOTA	1	£7,699.00	No
S752 LPY	200	1400	1998	23549	WHITE	ROVER	2	£3,450.00	Yes
SV03 YHP	MONDEO	1800	2003	37690	BLUE	FORD	1	£8,765.00	No

TASK 12: BONEHEART SURGERY

Student Information	REMEMBER:
In this task, you will create a database, then sort the information, query and database and produce a report.	You will create the file and enter some data into the database. You will need to be aware of all of the data types available and when to use each one appropriately. *Refer to pages 201 – 204, 206 – 208 and 211 – 214* for help with this task.

Creating a database and a report and preparing a safety statement

Scenario

The Practice Manager, Ida Foote, has asked you to create a database and enter some patient details. You decide to show Ida how useful databases are, so you produce queries and a report.

Activities

1   Open Access and create a new database named **Patients**

2   Create a new table and design it using the specification below:

Field	Data Type	Size
Patient number	Auto number	Long Integer
Surname	Text	20
First name	Text	30
Address	Text	50
Age	Number	Integer
Date of birth	Date/Time	Short Date
Gender	Text	10
Last appointment	Date/Time	Short date

3   **Save** your table naming it **Surgery - Your Name**

4   Make up information to include in the records for your table. You must enter at least **25** sets of details including at least 10 patients who are over 50 years old. **Be careful that the Age is correctly calculated in relation to the Date of Birth**.

5   **Sort** and **print** the table in **descending** order of **Age**.

6   Create two queries that filter the data using two selection criteria. For example, male patients over 50 and females patients born after 1980. **Save** the queries with suitable names (including **Your Name**) and print them.

7   Create a **report** using the following fields:

Surname; **First Name**; **Address**; **Date of Birth** and **Gender** in that order and with **Surname** sorted into **alphabetical orde**r (**A – Z**).

Use a **Report Title**:

Boneheart Surgery: Patient Details - Your name

8   **Print** the report and close the database.

9   Write a statement, of about 200 words, that describes how you can avoid risks to health when using computers for any length of time.

TASK 13: UWIN SOLICITORS

<table>
<tr>
<td>

Student Information

In this task, you will carry out some research and present the main facts in an email.

</td>
<td>

REMEMBER:

If you choose to use Google, the number of sites in the results can be restricted by selecting the **pages from the UK** search option.

You will need to have an email account.

Refer to pages 146 – 150 and 153 – 154 for help with this task.

</td>
</tr>
</table>

Researching information to produce a Word document; sending an email with an attachment

Scenario

You work for **Uwin Solicitors** and a client has requested information that relates to the **Sale of Goods Act 1994**. You are required to carry out some Internet research to locate appropriate information to enable you to answer the client's query.

Activities

> **Can my carpet fitter refuse to give me my money back if there is a fault in the carpet and I did not notice it until after they had gone? This was fitted last Friday.**
>
> **Mr I Loose**

1   Use an appropriate search engine such as **Google** or **Yahoo** and enter **Sale of Goods Act 1994** in the search box.

2   Locate sufficient information to enable you to answer the question. Print the relevant web pages.

3   Read through the information gathered and, using Word, create a document that summarises it into appropriate language for your client.

4   **Save** the information with a relevant filename.

5   **Print** the word-processed version for your client's file.

6   **Send** a copy of the document as a file attachment to the client by email. Include an appropriate message in the email. (Your tutor will act as the client.)

7   Hand in all your documents, including the research information.

TASK 14: FEEL GOOD FACTOR

<table>
<tr>
<td>

Student Information

In this task, you will be undertaking research, preparing an illustrated leaflet and attaching the document to an email.

</td>
<td>

REMEMBER:

You will need to be able to copy images and text from the Internet.

You will need to have an email account set up.

Check your work carefully for spelling and grammar.

Refer to pages 146 – 150 and 153 – 154 for help with this task.

</td>
</tr>
</table>

Researching information to produce a Word document; sending and receiving emails with file attachments

Scenario

The proprietor of **Feel Good Factor**, Kate Good, has asked you to put together a three to four page, A4 sized, Information Leaflet in order to advertise some of the treatments offered. You will use some of the information you find, including appropriate images, to put together a leaflet to give to potential clients.

Activities

1. Use the Internet and/or trade magazines to help you research the following range of treatments:

    *   Indian head massage;
    *   collagen facial treatments; and
    *   aromatherapy facial treatments.

    It would also be helpful to include appropriate images as illustrations.

2. **Copy** and **paste** the relevant information into an appropriate software program such as Word.

3. Develop the layout and structure that is appropriate for the purpose. Remember to include the Salon's name, **Feel Good Factor**, and make up suitable contact details.

4. **Save** the document with a suitable filename. (Remember to **save** the document regularly.)

5. **Print** a draft version and proof read it carefully, adding in pen any amendments you intend to make.

6. Edit the document, according to your notes, and when you are satisfied with the content and design send the document as an attachment to your tutor by **email**. Ask your tutor to reply to your email with his/her general comments.

7. **Print** a copy of the email you receive.

8. Amend your leaflet, if you wish, taking your tutor's comments into consideration.

9. **Print** and **save** a copy of the amended leaflet.

10. **Hand in** all printouts and research documents with your work.

TASK 15: FITNESS

Student Information	REMEMBER:
You will prepare a **Data Source** for a mail merge letter using some of the information shown in **Appendix 1**. You will also send mail merged letters to clients.	Read all instructions carefully. Use only the records that show members of the fitness centre that have **paid**. *Refer to pages 172 – 176* for help with this task.

Mail merge

Scenario

You are working as the marketing assistant for **Fitness**, a newly-opened gym. You are to send a welcome letter with a special offer to newly paid-up members, using your data source of clients.

Activities

Preparing a Mail Merge letter

1 Prepare the data source, **Appendix 1**, and the letter shown in **Appendix 2**.

> #### Remember
> Your data source should only contain those members who have **paid their membership**.

> #### Note
> Your data source contains extra Field Names. The data source you create for the letter will contain information related to **only** those <<Field Names>> used in the letter.

2 **Merge** the database with the letter and print the letters ready for despatch.

3 **Save all** documents before closing the software.

Title	First Name	Surname	Street	Town	County	Post Code	Tel No.	Email	Date Joined	M'ship Fee	Paid	Gender
Mrs	Alicia	French	1 Rockingham Rd	Alton	Derbyshire	AL1 9YU	01999 785478	a_french@ylassoo.co.uk	01/05/2006	£25.00	Yes	Female
Mr	Neil	Bromwell	23 Cedar Avenue	Midwest	Midlands	MW2 3OP	01888 123456	bromwell@nostampmail.com	10/06/2006	£45.00	Yes	Male
Miss	Kelly	Jacobs	41 The Grange	Brompton	Derbyshire	BR22 3LL	01777 999999	KJacobs@lassoo.co.uk	31/03/2006	£25.00	Yes	Female
Dr	David	Sundra	2 High Hill	Alton	Derbyshire	AL1 8RT	01999 987456	D.Sundra@mudkip.com	08/04/2006	£60.00	Yes	Male
Mr	James	Hunter	10 South Street	Brompton	Derbyshire	BR16 4KK	01777 333111	JH@uk-telecoms.co.uk	26/06/2006	£25.00	Yes	Male
Mrs	Claire	Ward	1 Rookup Villas	Alton	Derbyshire	AL4 1GH	01999 777858		29/06/2006	£45.00	No	Female
Mr	Robert	Muhoney	9 Burn View	Brompton	Derbyshire	BR7 3AB	01777 456789		03/07/2006	£60.00	No	Male

Today's date (dd/mm/yyyy)

<<Title>> <<First Name>> <<Surname>>
<<Street>>
<<Town>>
<<County>>
<<Post Code>>

Dear <<First Name>>

Re: A FITNESS Open Day

The Management of Fitness would like to extend a warm welcome to all its new members. We hope you are already feeling much fitter and healthier as a result of regular use of our high-quality facilities.

As a reward for your valued membership, we are very pleased to invite you to an Open Day on August 12th when you will be able to ask our expert fitness coaches for further advice, to enable you to get the very best from your membership.

Light refreshments will be available all day and these will be free of charge.

We look forward to seeing you.

Yours sincerely

A J Pressup
Manager

TASK 16: GREEN FINGERS

Student Information	REMEMBER:
In this task, you will be designing and completing a new quotation form. You will also be designing a diary sheet and completing it.	Save all of your documents regularly. Add your name to footers before printing. *Refer to pages 162 – 164, 177 – 179, 181 – 185, 187, 188 and 193 for help you with this task.*

Creating a spreadsheet and Word documents

Scenario

Mr Roy Green has an established garden landscaping partnership with his son Niall. At present he prepares quotations by hand, but does not keep a copy for his file. His son has now persuaded him that it is good practice to have a computerised form, which will improve the efficiency of the business. They hand you the task of creating a computer-based system and using it send out a quotation.

Activities

1   Examine the hand-written estimate that Roy hands to you (**Appendix 1**) that contains information from his visit to a prospective client.

2   Using Excel, design the quotation form that must include space for the **partnership's name and address**, **contact details**, **customers' name and address**, **description of the work**, **prices**, **VAT**, **total cost**.

It is suggested the columns you will need might be:

Job Description **Date of Visit** **Materials** **Quantities Needed**
Unit Price **Quantity** **Sub Total** **VAT**

Not necessarily in that order, but remember you are designing a spreadsheet that will make calculations so the information must allow for this.

You will be adding **Labour** to the cost, then **VAT** at **17.5%** to the cost of materials and labour, then providing a **Total** for the quotation.

Make full use of columns, headings, shading, ruling, merging cells, inserting an appropriate image to represent the company's logo. Make it attractive but professional. Simplicity of design is best.

3   **Save** and **print** a copy of the blank quotation.

4   Complete a quotation for Mrs Stone using the newly-designed form. Include formulae to provide all relevant totals.

5   **Save** the completed quotation as a separate file with a **new filename** because you want the original quotation to remain as the template.

6   **Print** two copies of the file saved in **Activity 4**, one showing the **values**, and one showing the **formulae**. **Make sure all the text is visible**.

7   Using Word, design a letter heading for the partnership, including the contact details and the logo.

(Continued)

8   **Print** and **save** the document.

9   Roy and Niall make a handwritten note of the work they have to complete for clients (*see* **Appendix 2**). Niall has decided the partnership needs a form that can be completed so the details can be recorded neatly and are easy to understand.

Using Word, design a diary sheet that will contain each month's work details, perhaps designed using a table. Add headings that will allow all the information in **Appendix 2** to be recorded. Add the company's logo and name and make use of shading, borders, etc. to enhance the design of the form.

10   **Save** the file and **print** a copy.

11   Use the information in **Appendix 1** to complete the form – remember to arrange in order of date and time. **Be consistent with the format of times**.

12   **Print** a copy of the completed form and **save** the file.

13   Write a statement of about 150 words identifying any problems you may have had with the equipment or the software packages you used in this task. If you did not experience any problems, say how you would have dealt with such things as the printer not working or paper getting jammed in the printer. How would you have dealt with problems saving, or retrieving your work? Did the software you were using allow you to do everything required of the task; if not, how did you overcome the problem?

Appendix 1

Green Fingers **The Paddock : Belfry Rise : Belway : Herefordshire : HR7 8NW**

Customer: Mrs B S Stone, 1 Heaton Drive, Belway, Herefordshire HR6 2CS
Tel - 364 577

Date of visit – 1 April 2006

Job Description
Rotavating garden plot at back of the property
Add Turf, Paving and Gravel

Materials required –
5 x 25 kg bags gravel   £50.00 per bag
8 sq metres paving   £20.00 sq metre
5 bags sand   £3.50 per bag   5 bags Cement £5.60 per bag
35 sq m turf   £2.50 per metre

Labour : 10 hours at £5 per hour
(Sub total these costs, then add VAT, then the overall Total).

Please prepare an estimate for Mrs Stone on the new Quotation form you are designing.

Thank you, Roy Green

Appointments for the month of (next month)

Mrs Geraldine Graham
2   Butterwick Lane, Belway
Tel 364762

Hedge trimming to front and side of property 5th 09:00am

Mr Jeremy Jenkins,
18 Elm Oval,
Belway   5th

11:30 am   364555
Relaying patio to rear of property

Mrs
Susan Kitson   7th   Relining garden pond
9am

Turnham Green House, Belway   364662

9 am   23rd   Mrs B S Stone, 1 Heaton Drive, Belway

Rotivating back garden, laying paving and turf and gravel

364577

Graham Painton 07754 456378

Gravelling drive area   14th
8:00

16 Mulberry Crescent, Belway

Mrs Diana Bradford   9th   1 pm   Hedge trimming

Ceddesfield House, The Green, Belway
364368

TASK 17: JUST MINUTE

Student Information

In this task, you will be creating, sorting and querying a database then producing a report from the query.

You will need to ask your tutor for the spreadsheet file called **Wrexham.xls** for this task.

You will also produce a set of mail merged letters.

REMEMBER:

Be consistent in the layout of text and numbers.

Make sure your printouts display all the information.

Prepare and print a screen shot of the database files you have created in this task.

Refer to pages 201 – 208, 172 – 176, 181 – 185, 187 – 188, 193, 211 – 214 and 145 for help with this task.

Creating, sorting and querying a database and producing a report; creating a spreadsheet; and producing mail merged letters

Scenario

You work for **Irlam and Daughter**, a model railway father and daughter partnership based in Swindon, Wiltshire. Trevithic Irlam acts as Secretary for the UK Model Railway Society and is putting together a database and letters related to forthcoming model railway exhibitions in the UK. Ireta Irlam, his daughter, has asked you to prepare a spreadsheet showing sales figures related to a recent exhibition the partnership attended.

Activities

Preparing the database

1   Have a look at the information contained in the table in **Appendix 1**.

2   Launch Access and create a new a database and give it a suitable filename for its purpose.

3   Design a new table containing fields and with appropriate data types for the information shown in **Appendix 1**.

4   **Save** the table, using an appropriate name (remember to include **Your Name**).

5   Enter the information shown in **Appendix 1**.

There have been some amendments and additions to the list. Make the following changes to the database:

6   The Isle of Wight venue has changed to the **Community Hall** with capacity of **58**.

   Another exhibition is to be held in **North Yorkshire**, in **Selby** in **Abbey Town Hall**, capacity **60**, on **1 September** for **one day**. The **Post Code** is **NY25 2MS**; the **Contact Name** is **Mrs Hilary Jordan**, who is the **Secretary**.

7   **Sort** the table into **alphabetical order** of **Location**.

8   Create a query that will **sort** the table into **ascending alphabetical order** of **County**.

9   **Save** the query with a suitable name (remember to include **Your Name**) and **print** a copy of the results.

10   **Query** the database, including all fields, to show those exhibitions that have a stand capacity of **over 55** being held for more than one day.

11   **Save** the query with a suitable name (include **Your Name**) and **Print** a copy.

12   Prepare a **report** from the query you saved in **Activity 11** to show the fields:

Location   Venue   Commencing Date   Number of Days (in that order).

13   Select a **report layout** and arrange the **Location** in **ascending** order.

14   Add a report title **Selected Locations - Your Name**

Make sure all the information can be seen and is displayed appropriately.

15   **Print** the report.

16   **Close** Access.

Preparing a mail merge letter

17   Prepare a letter for **mail merge** by entering some of the information from your database query into a **New Address List** in **Word**. . You will write the letter to the **Title** of the recipient because you know the names on your database might not be up to date. (The letter is shown in **Appendix 2**.)

18   **Print** a copy of the letter showing the **field names** and **letter text**.

19   **Merge** the letter with the information and **print** a copy of each letter.

20   **Close** Word.

Preparing the Spreadsheet

21   Launch Excel and open the spreadsheet called **Wrexham.xls**.

22   Insert a suitable company logo for the father and daughter partnership and position it appropriately. Adjust the column width, the height of rows and merge cells if necessary.

23   Display all the information consistently and format the cells appropriately.

24   Change the orientation of the heading **Catalogue No.** so it is consistent with the rest of the headings.

25   Insert formulae, replicating for **each** catalogue item in the following columns:

Purchase Price + VAT at 17.5%

Selling Price – which is (**Purchase Price + VAT @ 17.5%**) figure **plus** 12%

Value of Sales – which is (**Selling Price**) multiplied by (**No. Sold**)

Remaining Stock Levels – which is (**No. in Stock**) minus (**No. Sold**).

26   Provide a formulae for the **Total** of the following columns:

No. in Stock   No. Sold   Value of Sales   Remaining Stock Levels

27   Provide formula that will calculate: the **Average No. Sold**; the **Maximum No. Sold**; and the **Minimum No. Sold**. Placing the calculated figures in the **No. Sold** column. Make sure these calculations are displayed as **integer**.

28   Enhance the spreadsheet in any way you wish, and add a **custom foote**r.

29   **Print** two copies: one showing the **formulae** and one showing the **values**.

30   **Close** Excel.

31   Take a screen shot and print the image of the directory showing the filenames you created in this task.

County	Location	Venue	Post Code	Contact Name	Title	Number of Stands	Commencing Date	Number of Days
North Yorkshire	Helmsley	The White Swan	NY28 6JL	Miss Blyth Friar	Manager	60	18 November	2
Northamptonshire	Thrapston	Central Library Annexe	NH16 5GB	Mr Adrian Temple	Manager	45	21 August	3
Cambridgeshire	Catworth	Village Hall	CM11 5HP	Mrs Kate Grey	Organiser	60	12 November	3
Warwickshire	Dunchurch	The Grey Bear	WK14 3VK	Miss Ravinder Rahman	Functions Manager	50	8 Nov	2
Leicestershire	Earl Shilton	Community Hall	LC5 9BW	Mrs Laura Ferrier	Organiser	55	11 Oct	1
Durham	Darlington	Dolphin Centre	DH5 4BJ	Mr Simon Martins	Functions Manager	48	4 Dec	2
Cumbria	Furness	Town Hall	CM25 8DD	Mr Colin Furnace	Functions Organiser	60	8 October	2
Northumberland	Alnwick	Town Hall	NY31 6BS	Mr Harry Hanover	Chief Clerk	50	18 Oct	2
Borders	Coldstream	Coldstream Museum	BR11 9NZ	Mr Liam Scales	Manager	50	12 Sept	3
Aberdeenshire	Dyce	Town Hall	AD3 1FF	Mr Clive Reames	Chief Clerk	45	3 Nov	1
Wiltshire	Wootton Bassett	Lydiard Manor	WT5 8XW	Mr Troy Pawlett	Manager	55	24 October	2
Cleveland	Eaglescliffe	Preston Park Hall	TS18 5BZ	Miss Julie Earle	Events Manager	46	30 Oct	3
Gloucestershire	Patchway	Village Hall	GC6 9AW	Mr Allan Eichman	Organiser	45	3 Oct	2
Isle of Wight	Ventnor	The Cliff Hotel	WH4 8BL	Mr Peter Lynton	Manager	35	9 Sept	2
Jersey	St Aubin	St Aubin Hotel	JS3 12SN	Mr Linden Raif	Functions Manager	25	12 August	2
Isle of Man	Balladoole	Nautical Museum	IM15 9QJ	Mrs Laurel Kingfisher	Manager	35	16 August	2
Antrim	Lisburn	Trader Hall	LB3 5SG	Mrs Niamh Connell	Events Manager	67	1 Dec	2

UK Model Railway Exhibitions Society

7 Featherstone Place
SWINDON
Wiltshire
WT14 4YJ

Insert a suitable image and contact details please and display the letter heading as attractively as possible.

Today's date (dd/mm/yyyy)

<<Title>>
<<Venue>>
<<Address1>>
<<Address2>>
<<Post Code>>

Dear Sir or Madam

FORTHCOMING MODEL RAILWAY EXHIBITION

The purpose of this letter is to thank you for agreeing to let our Society hold its annual Model Railway Exhibition in the <<Venue>>.

I confirm the exhibition will take place on <<Number of Days>> commencing the <<Commencing date>>.

Enclosed is an Information Pack which gives details of the exhibitors and the opening times.

I thank you on behalf of the Society for your willingness to host this year's event.

Yours faithfully

Trevithic Irlam
Society Secretary

Enc

SAMPLE END ASSESSMENT

20 Multiple-choice questions

The following questions are multiple-choice. There is only one correct answer to each question.

Instructions

1   Choose whether you think the answer is A, B, C or D.

2   Ask your tutor for a copy of the answer grid (or download a copy from **www.lexden-publishing.co.uk/keyskills**).

3   Enter your answer on the marking grid at the end of the test.

4   Hand it to your tutor for marking.

An ICT Key Skills Level 2 External Assessment will consist of 40 questions and you will have **1 hour** to complete them.

How will you select your answers?

If you are sitting your End Assessment in paper format – not doing an online test – you will have to select one lettered answer for each numbered question. The answer sheet will be set in a similar way to the example below:

1   [a] [b] [c] [d]

2   [a] [b] [c] [d]

Make your choice by putting a **horizontal line** through the letter you think corresponds with the correct answer.

Use a pencil so you can alter your answer if you wish and take an eraser to allow you to change your mind about a response. Use an **HB pencil**, which is easier to erase. (If you make two responses for any one question, the question will be electronically marked as **incorrect**.)

Take a **black pen** into the exam room because you will have to sign the answer sheet.

Your tutor has 100 sample End Assessment questions and you will be given these when your tutor considers you are ready to practice the questions.

QUESTIONS

Questions 1 and 2 are about this database:

<table>
<tr><th colspan="6">Enrolment Details</th></tr>
<tr><th>ENROLMENT NO</th><th>SNAME</th><th>FNAME</th><th>AGE</th><th>GENDER</th><th>COURSE</th></tr>
<tr><td>SEC01</td><td>Cross</td><td>Ashley</td><td>17</td><td>F</td><td>Secretarial</td></tr>
<tr><td>MEC01</td><td>Johnson</td><td>David</td><td>17</td><td>M</td><td>Mechanics</td></tr>
<tr><td>MEC02</td><td>Smith</td><td>Jonathan</td><td>17</td><td>M</td><td>Mechanics</td></tr>
<tr><td>MEC03</td><td>Black</td><td>Gary</td><td>18</td><td>M</td><td>Mechanics</td></tr>
<tr><td>HSC01</td><td>Frost</td><td>Dorothy</td><td>18</td><td>F</td><td>Health & Social Care</td></tr>
<tr><td>BUL01</td><td>Dixon</td><td>Linda</td><td>19</td><td>F</td><td>Business & Law</td></tr>
<tr><td>BUL02</td><td>Jackson</td><td>Clare</td><td>17</td><td>F</td><td>Business & Law</td></tr>
<tr><td>SEC02</td><td>Burton</td><td>Melissa</td><td>19</td><td>F</td><td>Secretarial</td></tr>
<tr><td>MEC04</td><td>Carroll</td><td>Andrew</td><td>17</td><td>M</td><td>Mechanics</td></tr>
<tr><td>MEC05</td><td>Poole</td><td>James</td><td>18</td><td>M</td><td>Mechanics</td></tr>
<tr><td>HSC02</td><td>Taylor</td><td>Jayne</td><td>18</td><td>F</td><td>Health & Social Care</td></tr>
<tr><td>BUL03</td><td>Wright</td><td>Catherine</td><td>19</td><td>F</td><td>Business & Law</td></tr>
<tr><td>MEC06</td><td>Yorke</td><td>Steven</td><td>18</td><td>M</td><td>Mechanics</td></tr>
<tr><td>MEC07</td><td>Thompson</td><td>Raymond</td><td>19</td><td>M</td><td>Mechanics</td></tr>
</table>

1 The student who is enrolled on the Mechanics course and who is 19 is:

A   Steven Yorke

B   Raymond Thompson

C   Linda Dixon

D   Jonathan Smith

2 The search criterion to find Students enrolled on Mechanics courses is:

A   Enrol No. =Mechanics

B   Course =Mechanics

C   Enrol No. >=MEC01

D   Course=*MEC

Questions 3 to 5 are about this chart:

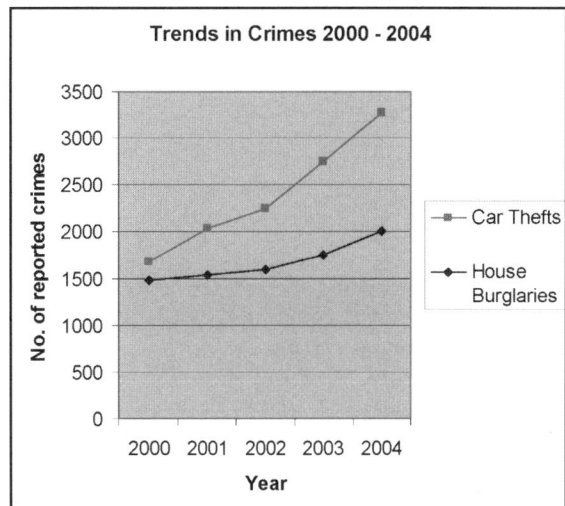

Trends in Crimes 2000 - 2004

3 The purpose of this chart is to:

A   explain the trends in crimes

B   promote police activity

C   compare types of crime

D   reduce crime

4   The legend explains:

A   which crimes are included in this line graph

B   which line style refers to which type of crime

C   which crimes have been committed

D   what crimes are committed every year

5   A method of presenting this data, not using graphical format, would be a:

A   bar chart

B   table

C   pie chart

D   column chart

6   Sometimes a user will forget a filename or not remember a folder in which a file was saved. In such a case the user should make use of:

A   a web browser

B   a search engine

C   find and replace

D   directory search tools

Questions 7 to 11 are about the spreadsheet at the bottom of the page.

Formulae have been used to calculate values in columns **D**, **G**, **H** and **I**.

7   The formula in cell D2 is:

A   =(B2/C2)

B   =B2*C2

C   =B2-C2

D   =B2+C2

8   The formula in cell **G2** is:

A   =F2/E2

B   =E2+F2

C   =E2*F2

D   =(F2*+D2)

9   The formula in **I2** to calculate the **% Profit** is:

A   =H2*G2

B   =H2/G2

C   =D2(H2*G2)

D   =SUM(H2:G2)

10   To present the data in each of the cells **B1:F1** in one line, rather then 2 lines, the user should:

A   increase the column width

B   increase the row height

C   use a larger font size

D   use a frame

11   The text in cells **B2 – B8**, **C2 – C8**, **D2 – D8**, **E2 – E8**, **G2 – G8** and **H2 – H8** would be more appropriately formatted using:

A   vertical alignment

B   left alignment

C   currency and align right

D   currency to 2 decimal places and align right

	A	B	C	D	E	F	G	H	I
1	Venue	Hall Cost	Artist Cost	Outgoings	Ticket Price	Tickets Sold	Income	Profit	% Profit
2	Cardiff	1200	2500	3700	10	1500	15000	11300	75%
3	Bristol	1600	2000	3600	5	800	4000	400	10%
4	Leeds	800	2500	3300	8	600	4800	1500	31%
5	York	75	2000	2075	7.5	850	6375	4300	67%
6	Lancaster	875	2000	2875	9.5	1200	11400	8525	75%
7	Glasgow	110	1500	1610	12	1600	19200	17590	92%
8	Dublin	950	1500	2450	10	1375	13750	11300	82%
9	Maximum Profit								92%

Questions 12 to 15 are about this line graph:

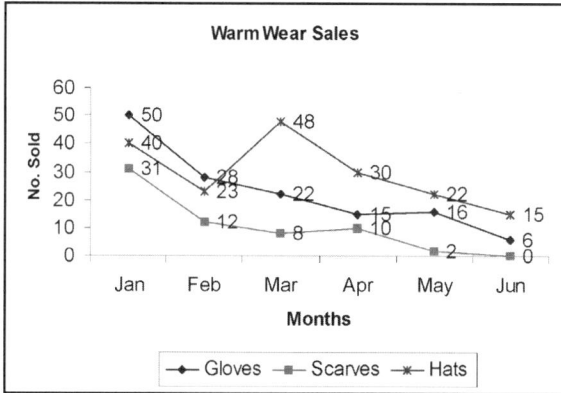

Warm Wear Sales

12 If the trends over the period May to June continue into July, which itjem or items will have to be discontinued because of lack of sales?

A   Gloves

B   Hats

C   Scarves

D   Scarves and Gloves

13 All sales decreased from:

A   Feb to Mar

B   Mar to June

C   May to June

D   Mar to May

14 A user wishes to make sure information stored on the computer is not lost if the computer breaks down. What should the user do?

A   back up all documents created and stored on the hard drive to CDs or DVDs

B   take a printout of every document saved

C   delete files once they have been printed

D   send important files to a friend for storage

15 The characters * and ? are sometimes used in search criteria as:

A   bookmarks

B   operators

C   wildcards

D   engines

Questions 16 and 17 are about searching the Internet

16 To simplify regular access to a website the user should:

A   save it

B   print it

C   refresh it

D   bookmark it

17 Using hotspots means:

A   following links set up on web pages

B   adding websites to the favorites list

C   using an advanced search in a search engine

D   using forward and back to move between web pages

Question 18 is about this chart:

Sales for January - May

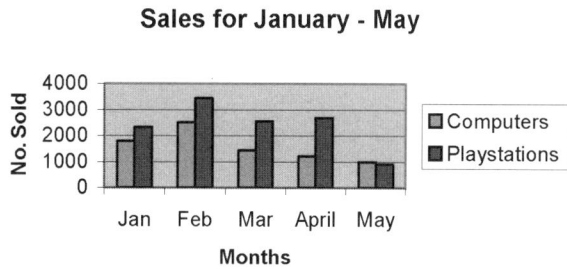

18 A legend is essential on this chart to show:

A   which bars represent which month

B   which bars represent which item

C   how many items have been sold

D   how many items were made

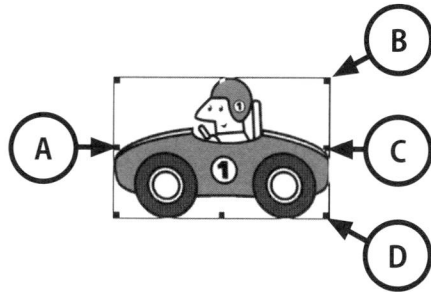

19 To adjust only the width of an image, the user should drag the handle at:

A   B or D

B   B

C   A or C

D   D

20 The user crops the image. This means the user has:

A   changed the colour of the image

B   resized the image

C   removed some of the original image

D   made one image from two images.

Index